Lydia Nash

A table book and introductory arithmetic

Lydia Nash

A table book and introductory arithmetic

ISBN/EAN: 9783741129865

Manufactured in Europe, USA, Canada, Australia, Japa

Cover: Foto ©berggeist007 / pixelio.de

Manufactured and distributed by brebook publishing software
(www.brebook.com)

Lydia Nash

A table book and introductory arithmetic

TABLE BOOK

AND

INTRODUCTORY ARITHMETIC.

———

BY L. NASH.

———

NEW YORK, CINCINNATI, CHICAGO:

BENZIGER BROTHERS,

PRINTERS TO THE   PUBLISHERS OF
HOLY APOSTOLIC SEE   BENZIGER'S MAGAZINE

PREFACE.

This little book has been compiled by a practical teacher to aid other teachers in drilling their pupils in the Simple Rules of Arithmetic.

It is believed that the Questions on the Tables will be found more complete than those of any similar manual. In the Exercises in Numeration and Notation, and in the Fundamental Rules, great pains has been taken to employ the cipher, always so puzzling to learners, in every possible combination with the significant figures.

As young children, for whom this book is intended, always depend upon their teachers for oral directions, rules and explanations have been deemed unnecessary.

ADDITION.

First Lesson.

1 and 0 are 1.	1 and 7 are how many?
1 and 1 are 2.	1 and 12 are how many?
1 and 2 are 3.	1 and 4 are how many?
1 and 3 are 4.	1 and 9 are how many?
1 and 4 are 5.	1 and 0 are how many?
1 and 5 are 6.	1 and 5 are how many?
1 and 6 are 7.	1 and 11 are how many?
1 and 7 are 8.	1 and 8 are how many?
1 and 8 are 9.	1 and 3 are how many?
1 and 9 are 10.	1 and 10 are how many?
1 and 10 are 11.	1 and 1 are how many?
1 and 11 are 12.	1 and 6 are how many?
1 and 12 are 13.	1 and 2 are how many?

Second Lesson.

2 and 0 are 2.	2 and 6 are how many?
2 and 1 are 3.	2 and 11 are how many?
2 and 2 are 4.	2 and 0 are how many?
2 and 3 are 5.	2 and 9 are how many?
2 and 4 are 6.	2 and 5 are how many?
2 and 5 are 7.	2 and 1 are how many?
2 and 6 are 8.	2 and 12 are how many?
2 and 7 are 9.	2 and 4 are how many?
2 and 8 are 10.	2 and 7 are how many?
2 and 9 are 11.	2 and 10 are how many?
2 and 10 are 12.	2 and 2 are how many?
2 and 11 are 13.	2 and 8 are how many?
2 and 12 are 14.	2 and 3 are how many?

Third Lesson.

3 and 0 are 3.	3 and 4 are how many?
3 and 1 are 4.	3 and 9 are how many?
3 and 2 are 5.	3 and 1 are how many?
3 and 3 are 6.	3 and 6 are how many?
3 and 4 are 7.	3 and 11 are how many?
3 and 5 are 8.	3 and 3 are how many?
3 and 6 are 9.	3 and 7 are how many?
3 and 7 are 10.	3 and 12 are how many?
3 and 8 are 11.	3 and 0 are how many?
3 and 9 are 12.	3 and 2 are how many?
3 and 10 are 13.	3 and 10 are how many?
3 and 11 are 14.	3 and 5 are how many?
3 and 12 are 15.	3 and 8 are how many?

Fourth Lesson.

4 and 0 are 4.	4 and 8 are how many?
4 and 1 are 5.	4 and 2 are how many?
4 and 2 are 6.	4 and 10 are how many?
4 and 3 are 7.	4 and 6 are how many?
4 and 4 are 8.	4 and 0 are how many?
4 and 5 are 9.	4 and 12 are how many?
4 and 6 are 10.	4 and 9 are how many?
4 and 7 are 11.	4 and 4 are how many?
4 and 8 are 12.	4 and 1 are how many?
4 and 9 are 13.	4 and 3 are how many?
4 and 10 are 14.	4 and 7 are how many?
4 and 11 are 15.	4 and 11 are how many?
4 and 12 are 16.	4 and 5 are how many?

Fifth Lesson.

5 and 0 are 5.	5 and 10 arc how many?
5 and 1 are 6.	5 and 4 are how many?
5 and 2 are 7.	5 and 8 are how many?
5 and 3 are 8.	5 and 1 are how many?
5 and 4 are 9.	5 and 5 are how many?
5 and 5 are 10.	5 and 12 are how many?
5 and 6 are 11.	5 and 2 are how many?
5 and 7 are 12.	5 and 7 are how many?
5 and 8 are 13.	5 and 0 are how many?
5 and 9 are 14.	5 and 9 are how many?
5 and 10 arc 15.	5 and 3 are how many?
5 and 11 are 16.	5 and 6 are how many?
5 and 12 are 17.	5 and 11 are how many?

Sixth Lesson.

6 and 0 are 6.	6 and 4 are how many?
6 and 1 are 7.	6 and 8 are how many?
6 and 2 are 8.	6 and 12 are how many?
6 and 3 are 9.	6 and 0 are how many?
6 and 4 are 10.	6 and 3 are how many?
6 and 5 are 11.	6 and 5 are how many?
6 and 6 are 12.	6 and 7 are how many?
6 and 7 are 13.	6 and 1 are how many?
6 and 8 are 14.	6 and 11 are how many?
6 and 9 are 15.	6 and 9 are how many?
6 and 10 are 16.	6 and 2 are how many?
6 and 11 are 17.	6 and 6 are how many?
6 and 12 are 18.	6 and 10 are how many?

Seventh Lesson.

7 and 0 are 7.	7 and 6 are how many?
7 and 1 are 8.	7 and 9 are how many?
7 and 2 are 9.	7 and 3 are how many?
7 and 3 are 10.	7 and 0 are how many?
7 and 4 are 11.	7 and 5 are how many?
7 and 5 are 12.	7 and 11 are how many?
7 and 6 are 13.	7 and 1 are how many?
7 and 7 are 14.	7 and 4 are how many?
7 and 8 are 15.	7 and 12 are how many?
7 and 9 are 16.	7 and 7 are how many?
7 and 10 are 17.	7 and 2 are how many?
7 and 11 are 18.	7 and 10 are how many?
7 and 12 are 19.	7 and 8 are how many?

Eighth Lesson.

8 and 0 are 8.	8 and 7 are how many?
8 and 1 are 9.	8 and 12 are how many?
8 and 2 are 10.	8 and 1 are how many?
8 and 3 are 11.	8 and 3 are how many?
8 and 4 are 12.	8 and 6 are how many?
8 and 5 are 13.	8 and 9 are how many?
8 and 6 are 14.	8 and 0 are how many?
8 and 7 are 15.	8 and 5 are how many?
8 and 8 are 16.	8 and 10 are how many?
8 and 9 are 17.	8 and 2 are how many?
8 and 10 are 18.	8 and 8 are how many?
8 and 11 are 19.	8 and 4 are how many?
8 and 12 are 20.	8 and 11 are how many?

Ninth Lesson.

9 and 0 are 9.	9 and 3 are how many?
9 and 1 are 10.	9 and 6 are how many?
9 and 2 are 11.	9 and 9 are how many?
9 and 3 are 12	9 and 4 are how many?
9 and 4 are 13.	9 and 2 are how many?
9 and 5 are 14.	9 and 8 are how many?
9 and 6 are 15.	9 and 12 are how many?
9 and 7 are 16.	9 and 0 are how many?
9 and 8 are 17.	9 and 5 are how many?
9 and 9 are 18.	9 and 11 are how many?
9 and 10 are 19.	9 and 1 are how many?
9 and 11 are 20.	9 and 10 are how many?
9 and 12 are 21.	9 and 7 are how many?

Tenth Lesson.

10 and 0 are 10.	10 and 10 are how many?
10 and 1 are 11.	10 and 7 are how many?
10 and 2 are 12.	10 and 3 are how many?
10 and 3 are 13.	10 and 12 are how many?
10 and 4 are 14.	10 and 0 are how many?
10 and 5 are 15.	10 and 9 are how many?
10 and 6 are 16.	10 and 6 are how many?
10 and 7 are 17.	10 and 2 are how many?
10 and 8 are 18.	10 and 11 are how many?
10 and 9 are 19.	10 and 8 are how many?
10 and 10 are 20.	10 and 1 are how many?
10 and 11 are 21.	10 and 5 are how many?
10 and 12 are 22.	10 and 4 are how many?

Eleventh Lesson.

11 and 0 are 11.	11 and 5 are how many?
11 and 1 are 12.	11 and 2 are how many?
11 and 2 are 13.	11 and 6 are how many?
11 and 3 are 14.	11 and 11 are how many?
11 and 4 are 15.	11 and 8 are how many?
11 and 5 are 16.	11 and 1 are how many?
11 and 6 are 17.	11 and 10 are how many?
11 and 7 are 18.	11 and 4 are how many?
11 and 8 are 19.	11 and 0 are how many?
11 and 9 are 20.	11 and 7 are how many?
11 and 10 are 21.	11 and 12 are how many?
11 and 11 are 22.	11 and 3 are how many?
11 and 12 are 23.	11 and 9 are how many?

Twelfth Lesson.

12 and 0 are 12.	12 and 8 are how many?
12 and 1 are 13.	12 and 12 are how many?
12 and 2 are 14	12 and 0 are how many?
12 and 3 are 15.	12 and 6 are how many?
12 and 4 are 16.	12 and 10 are how many?
12 and 5 are 17.	12 and 1 are how many?
12 and 6 are 18.	12 and 5 are how many?
12 and 7 are 19.	12 and 9 are how many?
12 and 8 are 20.	12 and 11 are how many?
12 and 9 are 21.	12 and 2 are how many?
12 and 10 are 22.	12 and 4 are how many?
12 and 11 are 23.	12 and 7 are how many?
12 and 12 are 24.	12 and 3 are how many?

Thirteenth Lesson.

2 and 4 are how many?	10 and 7 are how many?
6 and 5 are how many?	12 and 0 are how many?
8 and 10 are how many?	6 and 12 are how many?
7 and 4 are how many?	3 and 9 are how many?
12 and 2 are how many?	11 and 8 are how many?
8 and 7 are how many?	5 and 7 are how many?
3 and 4 are how many?	1 and 12 are how many?
2 and 12 are how many?	9 and 9 are how many?
11 and 11 are how many?	12 and 3 are how many?
9 and 0 are how many?	2 and 7 are how many?
1 and 8 are how many?	10 and 3 are how many?
4 and 5 are how many?	8 and 0 are how many?
5 and 12 are how many?	4 and 12 are how many?

Fourteenth Lesson.

7 and 6 are how many?	7 and 5 are how many?
12 and 1 are how many?	12 and 10 are how many?
2 and 8 are how many?	1 and 9 are how many?
11 and 1 are how many?	8 and 12 are how many?
8 and 9 are how many?	11 and 3 are how many?
1 and 6 are how many?	10 and 1 are how many?
3 and 5 are how many?	3 and 8 are how many?
10 and 12 are how many?	9 and 10 are how many?
5 and 9 are how many?	5 and 6 are how many?
4 and 0 are how many?	7 and 3 are how many?
9 and 3 are how many?	4 and 9 are how many?
6 and 11 are how many?	2 and 11 are how many?
9 and 9 are how many?	3 and 7 are how many?

Fifteenth Lesson.

3 and 6 are how many?   7 and 1 are how many?
1 and 0 are how many?   3 and 12 are how many?
7 and 9 are how many?   10 and 10 are how many?
11 and 4 are how many?   8 and 6 are how many?
6 and 7 are how many?   2 and 5 are how many?
10 and 9 are how many?   11 and 0 are how many?
2 and 2 are how many?   5 and 8 are how many?
8 and 3 are how many?   1 and 7 are how many?
5 and 0 are how many?   6 and 9 are how many?
9 and 12 are how many?   3 and 3 are how many?
1 and 5 are how many?   10 and 0 are how many?
12 and 4 are how many?   4 and 8 are how many?
4 and 7 are how many?   12 and 11 are how many?

Sixteenth Lesson.

6 and 8 are how many?   6 and 6 are how many?
11 and 10 are how many?   9 and 11 are how many'
9 and 6 are how many?   1 and 4 are how many'
8 and 5 are how many?   5 and 10 are how many?
12 and 9 are how many?   8 and 8 are how many?
6 and 0 are how many?   11 and 12 are how many?
1 and 2 are how many?   4 and 3 are how many?
4 and 6 are how many?   7 and 0 are how many?
10 and 8 are how many?   10 and 11 are how many?
5 and 1 are how many?   2 and 6 are how many?
2 and 10 are how many?   9 and 8 are how many?
3 and 0 are how many?   12 and 5 are how many?
7 and 8 are how many?   3 and 10 are how many?

Seventeenth Lesson.

1 and 1 are how many?
4 and 4 are how many?
6 and 2 are how many?
11 and 9 are how many?
5 and 5 are how many?
9 and 4 are how many?
7 and 12 are how many?
2 and 0 are how many?
12 and 6 are how many?
8 and 1 are how many?
3 and 11 are how many?
10 and 2 are how many?
1 and 11 are how many?

6 and 10 are how many?
11 and 2 are how many?
9 and 1 are how many?
5 and 11 are how many?
7 and 7 are how many?
1 and 3 are how many?
4 and 2 are how many?
12 and 8 are how many?
2 and 3 are how many?
8 and 11 are how many?
10 and 4 are how many?
3 and 1 are how many?
6 and 4 are how many?

Eighteenth Lesson.

9 and 7 are how many?
11 and 6 are how many?
5 and 4 are how many?
2 and 1 are how many?
7 and 11 are how many?
10 and 5 are how many?
4 and 11 are how many?
8 and 2 are how many?
12 and 7 are how many?
1 and 10 are how many?
3 and 2 are how many?
6 and 1 are how many?
9 and 5 are how many?

11 and 7 are how many?
4 and 1 are how many?
10 and 6 are how many?
5 and 2 are how many?
8 and 4 are how many?
9 and 2 are how many?
6 and 3 are how many?
11 and 5 are how many?
4 and 10 are how many?
7 and 2 are how many?
12 and 12 are how many?
5 and 3 are how many?
7 and 10 are how many?

SUBTRACTION.

Nineteenth Lesson.

1 from 1 leaves 0.	1 from 5 leaves how many?
1 from 2 leaves 1.	1 from 12 leaves how many?
1 from 3 leaves 2.	1 from 8 leaves how many?
1 from 4 leaves 3.	1 from 4 leaves how many?
1 from 5 leaves 4.	1 from 13 leaves how many?
1 from 6 leaves 5.	1 from 10 leaves how many?
1 from 7 leaves 6.	1 from 2 leaves how many?
1 from 8 leaves 7.	1 from 7 leaves how many?
1 from 9 leaves 8.	1 from 11 leaves how many?
1 from 10 leaves 9.	1 from 3 leaves how many?
1 from 11 leaves 10.	1 from 1 leaves how many?
1 from 12 leaves 11.	1 from 6 leaves how many?
1 from 13 leaves 12.	1 from 9 leaves how many?

Twentieth Lesson.

2 from 2 leaves 0.	2 from 6 leaves how many?
2 from 3 leaves 1.	2 from 11 leaves how many?
2 from 4 leaves 2.	2 from 2 leaves how many?
2 from 5 leaves 3.	2 from 14 leaves how many?
2 from 6 leaves 4.	2 from 5 leaves how many?
2 from 7 leaves 5.	2 from 8 leaves how many?
2 from 8 leaves 6.	2 from 3 leaves how many?
2 from 9 leaves 7.	2 from 12 leaves how many?
2 from 10 leaves 8.	2 from 7 leaves how many?
2 from 11 leaves 9.	2 from 10 leaves how many?
2 from 12 leaves 10.	2 from 4 leaves how many?
2 from 13 leaves 11.	2 from 13 leaves how many?
2 from 14 leaves 12.	2 from 9 leaves how many?

Tuesday. *Sept 27*

SUBTRACTION.

15

Twenty-first Lesson. *Study*

3 from 3 leaves 0.	3 from 12 leaves how many?
3 from 4 leaves 1.	3 from 9 leaves how many?
3 from 5 leaves 2.	3 from 6 leaves how many?
3 from 6 leaves 3.	3 from 3 leaves how many?
3 from 7 leaves 4.	3 from 15 leaves how many?
3 from 8 leaves 5.	3 from 7 leaves how many?
3 from 9 leaves 6.	3 from 11 leaves how many?
3 from 10 leaves 7.	3 from 14 leaves how many?
3 from 11 leaves 8.	3 from 5 leaves how many?
3 from 12 leaves 9.	3 from 10 leaves how many?
3 from 13 leaves 10.	3 from 4 leaves how many?
3 from 14 leaves 11.	3 from 8 leaves how many?
3 from 15 leaves 12.	3 from 13 leaves how many?

Twenty-second Lesson. *Failed Sept. 27*

4 from 4 leaves 0.	4 from 7 leaves how many?
4 from 5 leaves 1.	4 from 14 leaves how many?
4 from 6 leaves 2.	4 from 10 leaves how many?
4 from 7 leaves 3.	4 from 5 leaves how many?
4 from 8 leaves 4.	4 from 8 leaves how many?
4 from 9 leaves 5.	4 from 12 leaves how many?
4 from 10 leaves 6.	4 from 4 leaves how many?
4 from 11 leaves 7.	4 from 16 leaves how many?
4 from 12 leaves 8.	4 from 9 leaves how many?
4 from 13 leaves 9.	4 from 6 leaves how many!
4 from 14 leaves 10.	4 from 15 leaves how many!
4 from 15 leaves 11.	4 from 11 leaves how many!
4 from 16 leaves 12.	4 from 13 leaves how many!

Sept. 29 | 15 | 27 — Oct 4 (handwritten)

Study (handwritten)

SUBTRACTION.

Twenty-third Lesson.

5 from 5 leaves 0.	5 from 16 leaves how many?
5 from 6 leaves 1.	5 from 8 leaves how many?
5 from 7 leaves 2.	5 from 12 leaves how many?
5 from 8 leaves 3.	5 from 7 leaves how many?
5 from 9 leaves 4.	5 from 10 leaves how many?
5 from 10 leaves 5.	5 from 5 leaves how many?
5 from 11 leaves 6.	5 from 15 leaves how many?
5 from 12 leaves 7.	5 from 9 leaves how many?
5 from 13 leaves 8.	5 from 6 leaves how many?
5 from 14 leaves 9.	5 from 14 leaves how many?
5 from 15 leaves 10.	5 from 17 leaves how many?
5 from 16 leaves 11.	5 from 13 leaves how many?
5 from 17 leaves 12.	5 from 11 leaves how many?

Sept. 30 | 3rd | Oct (handwritten)   *Study Oct. 3rd* (handwritten)

Twenty-fourth Lesson.

6 from 6 leaves 0.	6 from 14 leaves how many?
6 from 7 leaves 1.	6 from 9 leaves how many?
6 from 8 leaves 2.	6 from 12 leaves how many?
6 from 9 leaves 3.	6 from 7 leaves how many?
6 from 10 leaves 4.	6 from 11 leaves how many?
6 from 11 leaves 5.	6 from 18 leaves how many?
6 from 12 leaves 6.	6 from 13 leaves how many?
6 from 13 leaves 7.	6 from 8 leaves how many?
6 from 14 leaves 8.	6 from 17 leaves how many?
6 from 15 leaves 9.	6 from 10 leaves how many?
6 from 16 leaves 10.	6 from 6 leaves how many?
6 from 17 leaves 11.	6 from 16 leaves how many?
6 from 18 leaves 12.	6 from 15 leaves how many?

7 from 7 leaves 0.	7 from 11 leaves how many?
7 from 8 leaves 1.	7 from 19 leaves how many?
7 from 9 leaves 2.	7 from 15 leaves how many?
7 from 10 leaves 3.	7 from 10 leaves how many?
7 from 11 leaves 4.	7 from 18 leaves how many?
7 from 12 leaves 5.	7 from 14 leaves how many?
7 from 13 leaves 6.	7 from 9 leaves how many?
7 from 14 leaves 7.	7 from 17 leaves how many?
7 from 15 leaves 8.	7 from 13 leaves how many?
7 from 16 leaves 9.	7 from 8 leaves how many?
7 from 17 leaves 10.	7 from 16 leaves how many?
7 from 18 leaves 11.	7 from 7 leaves how many?
7 from 19 leaves 12.	7 from 12 leaves how many?

Twenty-sixth Lesson.

8 from 8 leaves 0.	8 from 16 leaves how many?
8 from 9 leaves 1.	8 from 12 leaves how many?
8 from 10 leaves 2.	8 from 8 leaves how many?
8 from 11 leaves 3.	8 from 20 leaves how many?
8 from 12 leaves 4.	8 from 15 leaves how many?
8 from 13 leaves 5.	8 from 10 leaves how many?
8 from 14 leaves 6.	8 from 19 leaves how many?
8 from 15 leaves 7.	8 from 14 leaves how many?
8 from 16 leaves 8.	8 from 9 leaves how many?
8 from 17 leaves 9.	8 from 17 leaves how many?
8 from 18 leaves 10.	8 from 13 leaves how many?
8 from 19 leaves 11.	8 from 18 leaves how many?
8 from 20 leaves 12.	8 from 11 leaves how many?

Oct. 5,

Twenty-seventh Lesson.

9 from 9 leaves 0.	9 from 17 leaves how many?
9 from 10 leaves 1.	9 from 14 leaves how many?
9 from 11 leaves 2.	9 from 10 leaves how many?
9 from 12 leaves 3.	9 from 18 leaves how many?
9 from 13 leaves 4.	9 from 13 leaves how many?
9 from 14 leaves 5.	9 from 16 leaves how many?
9 from 15 leaves 6.	9 from 9 leaves how many?
9 from 16 leaves 7.	9 from 19 leaves how many?
9 from 17 leaves 8.	9 from 12 leaves how many?
9 from 18 leaves 9.	9 from 21 leaves how many?
9 from 19 leaves 10.	9 from 15 leaves how many?
9 from 20 leaves 11.	9 from 11 leaves how many?
9 from 21 leaves 12.	9 from 20 leaves how many?

Oct. 5,

X

Twenty-eighth Lesson.

10 from 10 leaves 0.	10 from 20 leaves how many?
10 from 11 leaves 1.	10 from 15 leaves how many?
10 from 12 leaves 2.	10 from 10 leaves how many?
10 from 13 leaves 3.	10 from 18 leaves how many?
10 from 14 leaves 4.	10 from 12 leaves how many?
10 from 15 leaves 5.	10 from 22 leaves how many?
10 from 16 leaves 6.	10 from 19 leaves how many?
10 from 17 leaves 7.	10 from 14 leaves how many?
10 from 18 leaves 8.	10 from 11 leaves how many?
10 from 19 leaves 9.	10 from 21 leaves how many?
10 from 20 leaves 10.	10 from 17 leaves how many?
10 from 21 leaves 11.	10 from 13 leaves how many?
10 from 22 leaves 12.	10 from 16 leaves how many.

SUBTRACTION.

Study *Oct 91?*

Please study these tables

Twenty-ninth Lesson.

11 from 11 leaves 0.	11 from 20 leaves how many?
11 from 12 leaves 1.	11 from 16 leaves how many?
11 from 13 leaves 2.	11 from 12 leaves how many?
11 from 14 leaves 3.	11 from 23 leaves how many?
11 from 15 leaves 4.	11 from 18 leaves how many?
11 from 16 leaves 5.	11 from 15 leaves how many?
11 from 17 leaves 6.	11 from 21 leaves how many?
11 from 18 leaves 7.	11 from 17 leaves how many?
11 from 19 leaves 8.	11 from 11 leaves how many?
11 from 20 leaves 9.	11 from 22 leaves how many?
11 from 21 leaves 10.	11 from 14 leaves how many?
11 from 22 leaves 11.	11 from 19 leaves how many?
11 from 23 leaves 12.	11 from 13 leaves how many?

Thirtieth Lesson.

Study *Oct 91*

12 from 12 leaves 0.	12 from 15 leaves how many?
12 from 13 leaves 1.	12 from 24 leaves how many?
12 from 14 leaves 2.	12 from 19 leaves how many?
12 from 15 leaves 3.	12 from 14 leaves how many?
12 from 16 leaves 4.	12 from 20 leaves how many?
12 from 17 leaves 5.	12 from 17 leaves how many?
12 from 18 leaves 6.	12 from 21 leaves how many?
12 from 19 leaves 7.	12 from 12 leaves how many?
12 from 20 leaves 8.	12 from 16 leaves how many?
12 from 21 leaves 9.	12 from 22 leaves how many?
12 from 22 leaves 10.	12 from 13 leaves how many?
12 from 23 leaves 11.	12 from 18 leaves how many?
12 from 24 leaves 12.	12 from 23 leaves how many?

Thirty-first Lesson.

2 from 6 leaves how many?   11 from 18 leaves how many?
7 from 17 leaves how many?   9 from 14 leaves how many?
5 from 8 leaves how many?   6 from 7 leaves how many?
12 from 12 leaves how many?   3 from 5 leaves how many?
7 from 9 leaves how many?   1 from 11 leaves how many?
4 from 14 leaves how many?   12 from 19 leaves how many?
11 from 16 leaves how many?   8 from 10 leaves how many?
6 from 9 leaves how many?   4 from 15 leaves how many?
5 from 7 leaves how many?   7 from 18 leaves how many?
9 from 11 leaves how many?   10 from 15 leaves how many?
8 from 12 leaves how many?   2 from 3 leaves how many?
10 from 16 leaves how many?   5 from 9 leaves how many?
4 from 5 leaves how many?   11 from 17 leaves how many?

Thirty-second Lesson.

9 from 16 leaves how many?   1 from 12 leaves how many?
1 from 4 leaves how many?   6 from 16 leaves how many?
6 from 10 leaves how many?   3 from 4 leaves how many?
3 from 14 leaves how many?   10 from 12 leaves how many?
10 from 14 leaves how many?   8 from 9 leaves how many?
8 from 19 leaves how many?   12 from 18 leaves how many?
2 from 5 leaves how many?   2 from 2 leaves how many?
12 from 20 leaves how many?   7 from 19 leaves how many?
4 from 6 leaves how many?   9 from 13 leaves how many?
7 from 14 leaves how many?   5 from 10 leaves how many?
5 from 16 leaves how many?   6 from 8 leaves how many?
9 from 10 leaves how many?   11 from 20 leaves how many?
11 from 13 leaves how many?   4 from 8 leaves how many?

Thirty-third Lesson.

1 from 2 leaves how many?	6 from 12 leaves how many?
3 from 13 leaves how many?	3 from 3 leaves how many?
12 from 17 leaves how many?	12 from 21 leaves how many?
2 from 8 leaves how many?	7 from 15 leaves how many?
9 from 17 leaves how many?	2 from 12 leaves how many?
10 from 21 leaves how many?	5 from 6 leaves how many?
7 from 7 leaves how many?	10 from 18 leaves how many?
11 from 23 leaves how many?	4 from 10 leaves how many?
4 from 7 leaves how many?	1 from 3 leaves how many?
8 from 16 leaves how many?	6 from 6 leaves how many?
5 from 15 leaves how many?	11 from 21 leaves how many?
1 from 5 leaves how many?	8 from 13 leaves how many?
9 from 20 leaves how many?	12 from 23 leaves how many?

Thirty-fourth Lesson.

6 from 14 leaves how many?	10 from 11 leaves how many?
3 from 7 leaves how many?	7 from 12 leaves how many?
2 from 13 leaves how many?	4 from 4 leaves how many?
7 from 10 leaves how many?	2 from 11 leaves how many?
5 from 11 leaves how many?	9 from 12 leaves how many?
4 from 13 leaves how many?	5 from 14 leaves how many?
9 from 19 leaves how many?	10 from 22 leaves how many?
3 from 11 leaves how many?	3 from 8 leaves how many?
8 from 20 leaves how many?	1 from 7 leaves how many?
11 from 14 leaves how many?	8 from 17 leaves how many?
1 from 10 leaves how many?	11 from 12 leaves how many?
6 from 17 leaves how many?	2 from 10 leaves how many?
12 from 22 leaves how many?	12 from 13 leaves how many?

Thirty-fifth Lesson.

7 from 13 leaves how many?   4 from 11 leaves how many?
4 from 12 leaves how many?   7 from 8 leaves how many?
10 from 10 leaves how many?   1 from 6 leaves how many?
9 from 15 leaves how many?   12 from 16 leaves how many?
3 from 6 leaves how many?   9 from 21 leaves how many?
1 from 8 leaves how many?   5 from 5 leave   y?
6 from 15 leaves how many?   8 from 11 leav   many?
5 from 13 leaves how many?   2 from 4 leaves how many?
11 from 11 leaves how many?   10 from 19 leaves how many?
2 from 7 leaves how many?   6 from 13 l   how many?
8 from 14 leaves how many?   11 from 15   s how many?
10 from 20 leaves how many?   3 from 9   ves how many?
3 from 15 leaves how many?   7 from 1 leaves how many

Thirty-sixth Lesson.

1 from 1 leaves how many?   12 from 14 leaves how many?
4 from 16 leaves how many?   5 from 17 leaves how many?
8 from 8 leaves how many?   4 from 9 leaves how many?
2 from 9 leaves how many?   11 from 22 leaves how many?
10 from 13 leaves how many?   2 from 14 leaves how many?
9 from 18 leaves how many?   9 from 9 leaves how many?
12 from 15 leaves how many?   3 from 7 leaves how many?
1 from 13 leaves how many?   8 from 15 leaves how many?
5 from 12 leaves how many?   12 from 24 leaves how many?
11 from 19 leaves how many?   1 from 9 le.
3 from 12 leaves how many?   7 from 11 leaves how many?
6 from 18 leaves how many?   8 from 18 leaves how many?
10 from 17 leaves how many?   6 from 11 leaves how many?

MULTIPLICATION.

Thirty-seventh Lesson.

Twice	0 is	0.	Twice	8	are how many?
Twice	1 is	2.	Twice	4	are how many?
Twice	2 are	4.*	Twice	12	are how many?
Twice	3 are	6.	Twice	0	is how many?
Twice	4 are	8.	Twice	7	are how many?
Twice	5 are	10.	Twice	10	are how many?
Twice	6 are	12.	Twice	2	are how many?
Twice	7 are	14.	Twice	6	are how many?
Twice	8 are	16.	Twice	11	are how many?
Twice	9 are	18.	Twice	1	is how many?
Twice	10 arc	20.	Twice	5	are how many?
Twice	11 are	22.	Twice	3	are how many?
Twice	12 are	24.	Twice	9	are how many?

Thirty-eighth Lesson.

3 times	0 is	0.	3 times	6	are how many?
3 times	1 is	3.	3 times	12	are how many?
3 times	2 are	6.	3 times	1	is how many?
3 times	3 are	9.	3 times	9	are how many?
3 times	4 are	12.	3 times	5	are how many?
3 times	5 are	15.	3 times	0	is how many?
3 times	6 are	18.	3 times	11	are how many?
3 times	7 are	21.	3 times	8	are how many?
3 times	8 are	24.	3 times	2	are how many?
3 times	9 are	27.	3 times	10	are how many?
3 times	10 are	30.	3 times	4	are how many?
3 times	11 are	33.	3 times	7	are how many?
3 times	12 are	36.	3 times	3	are how many?

* The best grammarians agree that in the Multiplication Table the t the word *times*, is the subject of the verb, consistency certainly requires that we should either say, Twice two *is* four, or, Two from four *leave* two, and Two in two *go* once: but it is still under consideration whether all the verbs shall be in the singular or all be made plural. Brown advocates the latter, Bullions the former. Meanwhile, I have adhered to the usual forms.

Thirty-ninth Lesson.

4 times 0 is 0.	4 times 9 are how many?
4 times 1 is 4.	4 times 3 are how many?
4 times 2 are 8.	4 times 5 are how many?
4 times 3 are 12.	4 times 11 are how many?
4 times 4 are 16.	4 times 0 is how many?
4 times 5 are 20.	4 times 7 are how many?
4 times 6 are 24.	4 times 2 are how many?
4 times 7 are 23.	4 times 12 are how many?
4 times 8 are 32.	4 times 4 are how many?
4 times 9 are 36.	4 times 10 are how many?
4 times 10 are 40.	4 times 1 is how many?
4 times 11 are 44.	4 times 8 are how many?
4 times 12 are 48.	4 times 6 are how many?

Fortieth Lesson.

5 times 0 is 0.	5 times 8 are how many?
5 times 1 is 5.	5 times 5 are how many?
5 times 2 are 10.	5 times 0 is how many?
5 times 3 are 15.	5 times 12 are how many?
5 times 4 are 20.	5 times 7 are how many?
5 times 5 are 25.	5 times 2 are how many?
5 times 6 are 30.	5 times 10 are how many?
5 times 7 are 35.	5 times 6 are how many?
5 times 8 are 40.	5 times 1 is how many?
5 times 9 are 45.	5 times 3 are how many?
5 times 10 are 50.	5 times 11 are how many?
5 times 11 are 55.	5 times 4 are how many?
5 times 12 are 90.	5 times 9 are how many?

Forty-first Lesson.

6 times 0 is 0. 6 times 6 are how many?
6 times 1 is 6.. 6 times 12 are how many?
6 times 2 are 12. 6 times 4 are how many?
6 times 3 are 18, 6 times 0 is how many?
6 times 4 are 24. 6 times 5 are how many?
6 times 5 are 30. 6 times 10 are how many?
6 times 6 are 36. 6 times 2 are how many?
6 times 7 are 42. 6 times 9 are how many?
6 times 8 are 48. 6 times 7 are how many?
6 times 9 are 54. 6 times 1 is how many?
6 times 10 are 60. 6 times 11 are how many?
6 times 11 are 66. 6 times 3 are how many?
6 times 12 are 72. 6 times 8 are how many?

Forty-second Lesson.

7 times 0 is 0. 7 times 4 are how many?
7 times 1 is 7. 7 times 9 are how many?
7 times 2 are 14. 7 times 11 are how many?
7 times 3 are 21. 7 times 1 is how many?
7 times 4 are 28. 7 times 5 are how many?
7 times 5 are 35. 7 times 8 are how many?
7 times 6 are 42. 7 times 10 are how many?
7 times 7 are 49. 7 times 0 is how many?
7 times 8 are 56. 7 times 3 are how many?
7 times 9 are 63. 7 times 12 are how many?
7 times 10 are 70. 7 times 2 are how many?
7 times 11 are 77. 7 times 7 are how many?
7 times 12 are 132. 7 times 6 are how many?

Forty-third Lesson.

8 times 0 is 0.	8 times 4 are how many?
8 times 1 is 8.	8 times 10 are how many?
8 times 2 are 16.	8 times 7 are how many?
8 times 3 are 24.	8 times 5 are how many?
8 times 4 are 32.	8 times 0 is how many?
8 times 5 are 40.	8 times 12 are how many?
8 times 6 are 48.	8 times 3 are how many?
8 times 7 are 56.	8 times 8 are how many?
8 times 8 are 64.	8 times 1 is how many?
8 times 9 are 72.	8 times 11 are how many?
8 times 10 are 80.	8 times 6 are how many?
8 times 11 are 88.	8 times 2 are how many?
8 times 12 are 96.	8 times 9 are how many?

Forty-fourth Lesson.

9 times 0 is 0.	9 times 5 are how many?
9 times 1 is 9.	9 times 8 are how many?
9 times 2 are 18.	9 times 0 is how many?
9 times 3 are 27.	9 times 11 are how many?
9 times 4 are 36.	9 times 4 are how many?
9 times 5 are 45.	9 times 9 are how many?
9 times 6 are 54.	9 times 2 are how many?
9 times 7 are 63.	9 times 12 are how many?
9 times 8 are 72.	9 times 7 are how many?
9 times 9 are 81.	9 times 1 is how many?
9 times 10 are 90.	9 times 10 are how many?
9 times 11 are 99.	times 3 are how many?
9 times 12 are 108.	9 × 9 are how many?

Forty-fifth Lesson.

10 times 0 is 0.	10 times 7 are how many?
10 times 1 is 10.	10 times 2 are how many?
10 times 2 are 20.	10 times 5 are how many?
10 times 3 are 30.	10 times 12 are how many?
10 times 4 are 40.	10 times 1 is how many?
10 times 5 are 50.	10 times 8 are how many?
10 times 6 are 60.	10 times 4 are how many?
10 times 7 are 70.	10 times 0 is how many?
10 times 8 are 80.	10 times 10 are how many?
10 times 9 are 90.	10 times 3 are how many?
10 times 10 are 100.	10 times 11 are how many?
10 times 11 are 110.	10 times 6 are how many?
10 times 12 are 120.	10 times 9 are how many?

Forty-sixth Lesson.

11 times 0 is 0.	11 times 5 are how many?
11 times 1 is 11.	11 times 11 are how many?
11 times 2 are 22.	11 times 4 are how many?
11 times 3 are 33.	11 times 1 is how many?
11 times 4 are 44.	11 times 12 are how many?
11 times 5 are 55.	11 times 3 are how many?
11 times 6 are 66.	11 times 6 are how many?
11 times 7 are 77.	11 times 0 is how many?
11 times 8 are 88.	11 times 8 are how many?
11 times 9 are 99.	11 times 10 are how many?
11 times 10 are 110.	11 times 2 are how many?
11 times 11 are 121.	11 times 9 are how many?
11 times 12 are 132.	11 times 7 are how many?

Forty-seventh Lesson.

12 times 0 is 0.	12 times 9 are how many?
12 times 1 is 12.	12 times 5 are how many?
12 times 2 are 24.	12 times 10 are how many?
12 times 3 are 36.	12 times 1 is how many?
12 times 4 are 48.	12 times 7 are how many?
12 times 5 are 60.	12 times 4 are how many?
12 times 6 are 72.	12 times 11 are how many?
12 times 7 are 84.	12 times 0 is how many?
12 times 8 are 96.	12 times 3 are how many?
12 times 9 are 108.	12 times 6 are how many?
12 times 10 are 120.	12 times 12 are how many?
12 times 11 are 132.	12 times 2 are how many?
12 times 12 are 144.	12 times 8 are how many?

Forty-eighth Lesson.

10 times 6 are how many?	3 times 11 are how many?
3 times 5 are how many?	10 times 9 are how many?
11 times 4 are how many?	11 times 2 are how many?
2 times 10 are how many?	7 times 4 are how many?
6 times 5 are how many?	12 times 9 are how many?
2 times 4 are how many?	5 times 2 are how many?
10 times 12 are how many?	8 times 7 are how many?
6 times 7 are how many?	4 times 6 are how many?
9 times 10 are how many?	9 times 1 is how many?
7 times 8 are how many?	11 times 6 are how many?
5 times 9 are how many?	2 times 2 are how many?
4 times 11 are how many?	6 times 3 are how many?
8 times 4 are how many?	5 times 11 are how many?

Forty-ninth Lesson.

5 times 4 are how many?	2 times 12 are how many?
8 times 3 are how many?	7 times 5 are how many?
12 times 2 are how many?	4 times 10 are how many?
5 times 1 is how many?	12 times 4 are how many?
9 times 3 are how many?	6 times 9 are how many?
6 times 11 are how many?	5 times 10 are how many?
7 times 7 are how many?	11 times 11 are how many?
3 times 4 are how many?	8 times 8 are how many?
10 times 11 are how many?	2 times 3 are how many?
9 times 12 are how many?	12 times 10 are how many!
4 times 2 are how many?	7 times 2 are how many!
11 times 5 are how many?	9 times 5 are how many!
2 times 9 are how many?	4 times 7 are how many!

Fiftieth Lesson.

11 times 12 are how many?	7 times 1 is how many!
7 times 10 are how many?	10 times 10 are how many?
8 times 11 are how many?	3 times 12 are how. many!
4 times 3 are how many?	8 times 6 are how many!
10 times 4 are how many?	10 times 1 is how many!
12 times 12 are how many?	3 times 8 are how many?
5 times 3 are how many?	2 times 6 are how many?
8 times 12 are how many?	9 times 11 are how many?
4 times 5 are how many?	12 times 6 are how many?
11 times 3 are how many?	11 times 1 is how many?
8 times 1 is how many?	2 times 5 are how many?
5 times 12 are how many?	9 times 8 are how many?
6 times 4 are how many?	6 times 10 are how many?

Fifty-first Lesson.

6 times 12 are how many?   4 times 9 are how many?
11 times 7 are how many?   10 times 2 are how many?
5 times 6 are how many?   5 times 7 are how many?
10 times 7 are how many?   12 times 8 are how many?
12 times 5 are how many?   2 times 11 are how many?
3 times 9 are how many?   4 times 4 are how many?
5 times 8 are how many?   7 times 11 are how many?
6 times 6 are how many?   3 times 3 are how many?
9 times 7 are how many?   6 times 2 are how many
7 times 3 are how many?   8 times 10 are how many.
9 times 2 are how many?   3 times 7 are how many?
11 times 8 are how many?   9 times 6 are how many?
3 times 10 are how many?   12 times 3 are how many?

Fifty-second Lesson.

11 times 9 are how many?   2 times 7 are how many?
6 times 8 are how many?   11 times 10 are how many?
10 times 5 are how many?   8 times 9 are how many?
4 times 1 is how many?   12 times 1 is how many?
12 times 11 are how many?   7 times 6 are how many?
5 times 5 are how many?   10 times 8 are how many?
7 times 9 are how many?   3 times 2 are how many?
8 times 2 are how many?   6 times 1 is how many?
10 times 3 are how many?   7 times 12 are how many?
12 times 7 are how many?   2 times 8 are how many?
4 times 8 are how many?   3 times 6 are how many?
9 times 4 are how many?   4 times 12 are how many?
8 times 5 are how many?   9 times 9 are how many?

DIVISION.

Fifty-third Lesson.

2 in 0 goes 0 times.				
2 in 2 goes once.				
2 in 4 goes twice.				
2 in 6 goes 3 times.				
2 in 8 goes 4 times.				
2 in 10 goes 5 times.				
2 in 12 goes 6 times.				
2 in 14 goes 7 times.				
2 in 16 goes 8 times.				
2 in 18 goes 9 times.				
2 in 20 goes 10 times.				
2 in 22 goes 11 times.				
2 in 24 goes 12 times.				

Fifty-fourth Lesson.

3 in 0 goes 0 times.				
3 in 3 goes once.				
3 in 6 goes twice.				
3 in 9 goes 3 times.				
3 in 12 goes 4 times.				
3 in 15 goes 5 times.				
3 in 18 goes 6 times.				
3 in 21 goes 7 times.				
3 in 24 goes 8 times.				
3 in 27 goes 9 times.				
3 in 30 goes 10 times.				
3 in 33 goes 11 times.				
3 in 36 goes 12 times.				

Fifty-first Lesson.

6 times 12 are how many? 4 times 9 are how many?
11 times 7 are how many? 10 times 2 are how many?
5 times 6 are how many? 5 times 7 are how many?
10 times 7 are how many? 12 times 8 are how many?
12 times 5 are how many? 2 times 11 are how many?
3 times 9 are how many? 4 times 4 are how many?
5 times 8 are how many? 7 times 11 are how many?
6 times 6 are how many? 3 times 3 are how many?
9 times 7 are how many? 6 times 2 are how many
7 times 3 are how many? 8 times 10 are how many.
9 times 2 are how many? 3 times 7 are how many?
11 times 8 are how many? 9 times 6 are how many?
3 times 10 are how many? 12 times 3 are how many?

Fifty-second Lesson.

11 times 9 are how many? 2 times 7 are how many?
6 times 8 are how many? 11 times 10 are how many?
10 times 5 are how many? 8 times 9 are how many?
4 times 1 is how many? 12 times 1 is how many?
12 times 11 are how many? 7 times 6 are how many?
5 times 5 are how many? 10 times 8 are how many?
7 times 9 are how many? 3 times 2 are how many?
8 times 2 are how many? 6 times 1 is how many?
10 times 3 are how many? 7 times 12 are how many?
12 times 7 are how many? 2 times 8 are how many?
4 times 8 are how many? 3 times 6 are how many?
9 times 4 are how many? 4 times 12 are how many?
8 times 5 are how many? 9 times 9 are how many?

DIVISION.

Fifty-third Lesson.

2 in 0 goes 0 times.	2 in 8 how many times?
2 in 2 goes once.	2 in 12 how many times?
2 in 4 goes twice.	2 in 16 how many times?
2 in 6 goes 3 times.	2 in 20 how many times?
2 in 8 goes 4 times.	2 in 24 how many times?
2 in 10 goes 5 times.	2 in 6 how many times?
2 in 12 goes 6 times.	2 in 2 how many times?
2 in 14 goes 7 times.	2 in 10 how many times?
2 in 16 goes 8 times.	2 in 14 how many times?
2 in 18 goes 9 times.	2 in 22 how many times?
2 in 20 goes 10 times.	2 in 4 how many times?
2 in 22 goes 11 times.	2 in 0 how many times?
2 in 24 goes 12 times.	2 in 18 how many times?

Fifty-fourth Lesson.

3 in 0 goes 0 times.	3 in 24 how many times?
3 in 3 goes once.	3 in 33 how many times?
3 in 6 goes twice.	3 in 12 how many times?
3 in 9 goes 3 times.	3 in 0 how many times?
3 in 12 goes 4 times.	3 in 27 how many times?
3 in 15 goes 5 times.	3 in 36 how many times?
3 in 18 goes 6 times.	3 in 18 how many times?
3 in 21 goes 7 times.	3 in 6 how many times?
3 in 24 goes 8 times.	3 in 30 how many times?
3 in 27 goes 9 times.	3 in 3 how many times?
3 in 30 goes 10 times.	3 in 15 how many times?
3 in 33 goes 11 times.	3 in 9 how many times?
3 in 36 goes 12 times.	3 in 21 how many times?

Fifty-fifth Lesson.

4 in 0 goes 0 times.	4 in 20 how many times?
4 in 4 goes once.	4 in 32 how many times?
4 in 8 goes twice.	4 in 40 how many times?
4 in 12 goes 3 times.	4 in 0 how many times?
4 in 16 goes 4 times.	4 in 16 how many times?
4 in 20 goes 5 times.	4 in 48 how many times?
4 in 24 goes 6 times.	4 in 8 how many times?
4 in 28 goes 7 times.	4 in 36 how many times?
4 in 32 goes 8 times.	4 in 44 how many times?
4 in 36 goes 9 times.	4 in 4 how many times?
4 in 40 goes 10 times.	4 in 24 how many times?
4 in 44 goes 11 times.	4 in 12 how many times?
4 in 48 goes 12 times.	4 in 28 how many times?

Fifty-sixth Lesson.

5 in 0 goes 0 times.	5 in 45 how many times
5 in 5 goes once.	5 in 20 how many times
5 in 10 goes twice.	5 in 30 how many times
5 in 15 goes 3 times.	5 in 5 how many times
5 in 20 goes 4 times.	5 in 55 how many times
5 in 25 goes 5 times.	5 in 25 how many times
5 in 30 goes 6 times.	5 in 0 how many times
5 in 35 goes 7 times.	5 in 60 how many times
5 in 40 goes 8 times.	5 in 10 how many times
5 in 45 goes 9 times.	5 in 40 how many times
5 in 50 goes 10 times.	5 in 15 how many times
5 in 55 goes 11 times.	5 in 50 how many times
5 in 60 goes 12 times.	5 in 35 how many times

Sept 24 *study for Sept 2?*

Fifty-seventh Lesson.

6 in 0 goes 0 times.	6 in 42 how many times?
6 in 6 goes once.	6 in 54 how many times?
6 in 12 goes twice.	6 in 66 how many times?
6 in 18 goes 3 times.	6 in 6 how many times?
6 in 24 goes 4 times.	6 in 36 how many times?
6 in 30 goes 5 times.	6 in 72 how many times?
6 in 36 goes 6 times.	6 in 0 how many times?
6 in 42 goes 7 times.	6 in 18 how many times?
6 in 48 goes 8 times.	6 in 60 how many times?
6 in 54 goes 9 times.	6 in 24 how many times?
6 in 60 goes 10 times.	6 in 12 how many times?
6 in 66 goes 11 times.	6 in 30 how many times?
6 in 72 goes 12 times.	6 in 48 how many times?

Fifty-eighth Lesson. *study*

7 in 0 goes 0 times.	7 in 49 how many times?
7 in 7 goes once.	7 in 84 how many times?
7 in 14 goes twice.	7 in 63 how many times?
7 in 21 goes 3 times.	7 in 0 how many times?
7 in 28 goes 4 times.	7 in 21 how many times?
7 in 35 goes 5 times.	7 in 70 how many times?
7 in 42 goes 6 times.	7 in 35 how many times?
7 in 49 goes 7 times.	7 in 7 how many times?
7 in 56 goes 8 times.	7 in 28 how many times?
7 in 63 goes 9 times.	7 in 77 how many times?
7 in 70 goes 10 times.	7 in 56 how many times?
7 in 77 goes 11 times.	7 in 14 how many times?
7 in 84 goes 12 times.	7 in 42 how many times?

Sept. 29   *Failed*   *Sept. 26*

Sept. 30

Fifty-ninth Lesson.

8 in	0	goes	0 times.	8 in	16	how many times?
8 in	8	goes	once.	8 in	88	how many times?
8 in	16	goes	twice.	8 in	64	how many times?
8 in	24	goes	3 times.	8 in	48	how many times?
8 in	32	goes	4 times.	8 in	0	how many times?
8 in	40	goes	5 times.	8 in	80	how many times?
8 in	48	goes	6 times.	8 in	56	how many times?
8 in	56	goes	7 times.	8 in	40	how many times?
8 in	64	goes	8 times.	8 in	72	how many times?
8 in	72	goes	9 times.	8 in	96	how many times?
8 in	80	goes	10 times.	8 in	8	how many times?
8 in	88	goes	11 times.	8 in	32	how many times?
8 in	96	goes	12 times.	8 in	24	how many times?

Oct 3

Sept. 29 | *27 study*

Sixtieth Lesson.

9 in	0	goes	0 times.	9 in	81	how many times?
9 in	9	goes	once.	9 in	63	how many times?
9 in	18	goes	twice.	9 in	45	how many times?
9 in	27	goes	3 times.	9 in	108	how many times?
9 in	36	goes	4 times.	9 in	9	how many times?
9 in	45	goes	5 times.	9 in	54	how many times?
9 in	54	goes	6 times.	9 in	27	how many times?
9 in	63	goes	7 times.	9 in	99	how many times?
9 in	72	goes	8 times.	9 in	0	how many times?
9 in	81	goes	9 times.	9 in	72	how many times?
9 in	90	goes	10 times.	9 in	36	how many times?
9 in	99	goes	11 times.	9 in	90	how many times?
9 in	108	goes	12 times.	9 in	18	how many times?

Failed

˜st Lesson.

10 in	0	goes	ʊ	in	50	how many times?
10 in	10	goes	once.		90	how many times?
10 in	20	goes	twice.		˜0	how many times?
10 in	30	goes	3 times.			how many times?
10 in	40	goes	4 times.	˥		v many times?
10 in	50	goes	5 times.	10 ˩.		˥any times?
10 in	60	goes	6 times.	10 in .		˥y times?
10 in	70	goes	7 times.	10 in ˩		times?
10 in	80	goes	8 times.	10 in 40 ˩.		
10 in	90	goes	9 times.	10 in 110 hoᵥ		
10 in	100	goes	10 times.	10 in 20 how ˩.		
10 in	110	goes	11 times.	10 in 80 how ma˩		
10 in	120	goes	12 times.	10 in 70 how many		

Sixty-second Lesson.

11 in	0	goes	0 times.	11 in	66	how many times?
11 in	11	goes	once.	11 in	99	how many times?
11 in	22	goes	twice.	11 in	132	how many times?
11 in	33	goes	3 times.	11 in	0	how many times?
11 in	44	goes	4 times.	11 in	88	how many times?
11 in	55	goes	5 times.	11 in	44	how many times?
11 in	66	goes	6 times.	11 in	11	how many times?
11 in	77	goes	7 times.	11 in	77	how many times?
11 in	88	goes	8 times.	11 in	121	how many times?
11 in	99	goes	9 times.	11 in	22	how many times?
11 in	110	goes	10 times.	11 in	55	how many times?
11 in	121	goes	11 times.	11 in	33	how many times?
11 in	132	goes	12 times.	11 in	110	how many times?

Sixty-third Lesson.

12 in 0 goes 0 times.	12 in 120 how many times
12 in 12 goes once.	12 in 36 how many times?
12 in 24 goes twice.	12 in 72 how many times?
12 in 36 goes 3 times.	12 in 12 how many times?
12 in 48 goes 4 times	12 in 48 how many times?
12 in 60 goes 5 times.	12 in 144 how many times?
12 in 72 goes 6 times.	12 in 0 how many times?
12 in 84 goes 7 times.	12 in 96 how many times?
12 in 96 goes 8 times.	12 in 132 how many times?
12 in 108 goes 9 times.	12 in 24 how many times?
12 in 120 goes 10 times.	12 in 60 how many times?
12 in 132 goes 11 times.	12 in 108 how many times?
12 in 144 goes 12 times.	12 in 84 how many times?

Sixty-fourth Lesson.

9 in 36 how many times?	4 in 20 how many times?
11 in 66 how many times?	5 in 60 how many times?
12 in 144 how many times?	10 in 70 how many times?
2 in 8 how many times?	12 in 84 how many times?
6 in 30 how many times?	6 in 72 how many times?
8 in 80 how many times?	3 in 27 how many times?
7 in 28 how many times?	11 in 88 how many times?
12 in 24 how many times?	5 in 35 how many times?
8 in 56 how many times?	12 in 36 how many times?
3 in 12 how many times?	9 in 81 how many times?
2 in 24 how many times?	10 in 90 how many times?
11 in 121 how many times?	2 in 14 how many times?
9 in 45 how many times?	8 in 16 how many times?

Sixty-fifth Lesson.

4 in 48 how many times?	3 in 36 how many times?
3 in 18 how many times?	10 in 100 how many times?
7 in 63 how many times?	8 in 48 how many times?
11 in 44 how many times?	2 in 10 how many times?
6 in 42 how many times?	11 in 77 how many times?
10 in 90 how many times?	5 in 40 how many times?
2 in 4 how many times?	6 in 54 how many times?
8 in 24 how many times?	3 in. 9 how many times?
5 in 10 how many times?	12 in 12 how many times?
9 in 108 how many times?	9 in 54 how many times?
12 in 48 how many times?	10 in 50 how many times!
7 in 7 how many times?	4 in 32 how many times?
4 in 28 how many times?	7 in 42 how many times?

Sixty-sixth Lesson.

2 in 16 how many times?	8 in 96 how many times?
11 in 11 how many times?	11 in 33 how many times?
8 in 72 how many times?	10 in 10 how many times?
7 in 70 how many times?	3 in 24 how many times?
3 in 15 how many times?	9 in 90 how many times?
10 in 120 how many times?	5 in 30 how many times?
5 in 45 how many times?	7 in 21 how many times?
4 in 4 how many times?	4 in 36 how many times?
9 in 27 how many times?	2 in 22 how many times?
6 in 66 how many times?	3 in 21 how many times?
2 in 18 how many times?	6 in 48 how many times?
7 in 35 how many times?	12 in 132 how many times?
2 in 120 how many times?	8 in 40 how many times?

Sixty-seventh Lesson.

11 in 110 how many times?	11 in 132 how many times?
6 in 6 how many times?	4 in 12 how many times?
4 in 24 how many times?	8 in 32 how many times?
10 in 80 how many times?	10 in 110 how many times?
5 in 5 how many times?	2 in 12 how many times?
2 in 20 how many times?	9 in 72 how many times?
7 in 56 how many times?	12 in 60 how many times?
12 in 108 how many times?	3 in 30 how many times?
3 in 6 how many times?	4 in 16 how many times?
6 in 36 how many times?	6 in 12 how many times?
9 in 99 how many times?	11 in 99 how many times?
5 in 50 how many times?	5 in 25 how many times?
8 in 64 how many times?	7 in 84 how many times?

Sixty-eighth Lesson.

4 in 44 how many times?	5 in 55 how many times?
12 in 72 how many times?	7 in 49 how many times?
8 in 8 how many times?	4 in 8 how many times?
3 in 33 how many times?	12 in 96 how many times?
10 in 20 how many times?	2 in 6 how many times?
9 in 18 how many times?	8 in 88 how many times?
11 in 55 how many times?	10 in 40 how many times?
4 in 40 how many times?	6 in 24 how many times?
7 in 14 how many times?	9 in 63 how many times?
5 in 15 how many times?	10 in 60 how many times?
9 in 9 how many times?	5 in 20 how many times?
6 in 90 how many times?	6 in 18 how many times?
11 in 12 how many times?	7 in 77 how many times?

NUMERATION and NOTATION.

NUMERATION is the art of reading numbers.

NOTATION is the art of writing numbers.

In order to read numbers we separate them into periods of three figures each, beginning to count at the right hand.

There must be three figures in every period except the one on the left hand, which may contain less than three but not more.

The names of the first five periods from right to left are, as follows:

Units, Thousands, Millions, Billions, Trillions.

From left to right:

Trillions, Billions, Millions, Thousands, Units.

Read the following :*

1	100	9	52000
2	605	10	60100
3	1000	11	90704
4	9010	12	100000
5	8700	13	250600
6	4009	14	905804
7	10000	15	400008
8	20001	16	706002

* The pupil is expected to copy these numbers upon his slate and point them off, numerate them, and mention the periods, before reading them.

17	. . .800090	48	800000200
18	300700	49	1000000000
19	'.401000	50	6000000006
20	1000000	51	7700007000
21	6002000	52	4200000000
22	8203000	53	1000000000
23	7060000	54	80000000080
24	4000008	55	75000000700
25	2090070	56	90000008000
26	9800200	57	60245070000
27	3070009	58	100000000000
28	5309020	59	608005004002
29	6008525	60	400400400400
30	10000000	61	100000080000
31	80080080	62	800040624000
32	60006000	63	100000000000
33	25400000	64	860000000000
34	40400049	65	4004802090700
35	30006001	66	6000200030002
36	90085700	67	4320085690024
37	50000600	68	5000000060000
38	10000004	69	10000000000000
39	100000000	70	20060001008010
40	700000080	71	40110480974200
41	600800900	72	30008254009240
42	204600000	73	70090002000750
43	110040000	74	60002000900600
44	200000004	75	100000000000000
45	750000805	76	101009082700265
46	400310060	77	700400006348100
47	300500000	78	600020001000825

Write the following in figures:

1. Six thousand.
2. Three millions five hundred thousand six hundred.
3. Twenty millions thirty thousand and forty.
4. Sixty billions two hundred forty-five millions and seventy.
5. Nine trillions.
6. Forty millions thirty-four thousand six hundred and twenty-five.
7. Three hundred millions eight hundred thousand nine hundred and five.
8. Four hundred trillions six billions one hundred eighty-nine millions seven hundred sixty thousand eight hundred and four.
9. Five millions.
10. Four millions five thousand.
11. Sixty thousand.
12. Twenty thousand and five.
13. Two hundred thousand.
14. Three hundred millions thirty-five thousand and six.
15. Five billions.
16. Seven hundred millions.
17. Ten trillions and eighty.
18. Forty-two billions two hundred.
19. Eighty trillions seventy billions two millions nine thousand and ten.
20. Three thousand and seven.

21. Forty thousand two hundred and eight.

22. Eight millions twenty thousand and sixty.

23. Four millions six hundred eight thousand and fifty-seven.

24. Seven millions two thousand eight hundred and twenty-one.

25. Sixty millions.

26. Thirty millions and two thousand.

27. Eighty-four millions seven hundred thousand.

28. Ten millions two hundred.

29. Two hundred millions.

30. Seven hundred seventy-five millions four hundred and sixty-two.

31. Four thousand and fifty.

32. Two thousand five hundred.

33. Eight hundred thousand three hundred and fourteen.

34. Six millions and five.

35. Three hundred billions six hundred thousand and fifty-two.

36. Four billions three millions two thousand and ten.

37. Eighteen trillions and six hundred.

38. Two trillions nine billions seventy millions four hundred six thousand and eighty.

39. Four hundred trillions ten billions eight millions nine hundred and sixty-five.

In the ROMAN method of Notation letters are used instead of figures.

One is written..........I	20 is written.......XX
2 is written.... II	21 is written......XXI
3 is written..........III	30 is written.....XXX
4 is written..........IV	40 is written......XL
5 is written.......... V	50 is written........L
6 is written..........VI	60 is written......LX
7 is written.........VII	70 is written.....LXX
8 is written....... VIII	80 is written....LXXX
9 is written..........IX	90 is written......XC
10 is written.......... X	100 is written.........C
11 is written..........XI	200 is written...... CC
12 is written... XII	300 is written......CCC
13 is written........ XIII	400 is written.... CCCC
14 is written........ XIV	500 is written........D
15 is written......... XV	600 is written...... DC
16 is written........ XVI	700 is written.....DCC
17 is written....... XVII	800 is written.... DCCC
18 is written...... XVIII	900 is written...DCCCC
19 is written........ XIX	1000 is written........ M

What does XXXIV stand for? LV? LXXXVIII? XLVI? XXIV? LXVII? XCIII? LXXIX? CLVIII? MD? DCXLII? DCCCCLXXXVIII? CCCXXIII? CCXVI? MDCCXV? CCCCLXXVI? DCCXIX? MCIX? DX? MLXVI? MCCCCXCII? MDCVIF? MDCXX? MDCCLXXVI? MDCCCLXI? MDCCCLXXVIII?

EXERCISES FOR THE SLATE.

ADDITION.

ADDITION is the process of collecting several numbers into one.

The answer in Addition is called the Sum or Amount.

	(1)	(2)
Add	1365418321	2140658214
To	4514261323	6539130514
A.	5879679644	

(3)	(4)	(5)
846921765	43687195	9765487
982765497	87696769	2798769

(6)	(7)	(8)
98764829	18926478	9647621
76596473	97698462	6589765

(9)	(10)	(11)
765987642	29831576	182769
876548976	15278654	376145
214372465	36890749	823762

(12)	(13)	(14)
84269158	87698421	84769872
78987692	43987687	65476587
41768706 ·	69476876	76189824

(15)	(16)	(17)
689765143	81468916	876982437
871468721	19876897	698765146
659876984	65410762	875649825
847698765	43291654	737196872

(18)	(19)	(20)
765143261	2768715	14691821
947876542	7692159	73082754
176189765	1087608	61538726
876789648	2462758	95029179

(21)	(22)	(23)
659876214	23469821	1476521
371698763	87698765	8769876
476918924	40372659	4698259
371689743	81427695	7367824
817916985	41876892	6915762

(24)	(25)	(26)
84698275	36918243	2346987
69176872	67827465	3468295
91587643	91276876	4376258
47698765	46912918	7165876
74192765	76143876	9347283

(27)	(28)	(29)
43769872	46067652	14569876
85976514	87698769	46967692
67136982	76240982	71876914
41378497	36159876	63215765
71876914	19876214	47869582
64369870	91768765	32154398

(30)	(31)	(32)
43789761	14691876	48769154
54891695	87643762	87689762
21437698	34187695	41569874
46918769	41658768	37168724
76514376	97691875	56073156
19876543	67629869	97640982

(33)	(34)	(35)
14876918	46987694	76514987
49165465	87651469	82497687
21371598	12987654	59876876
76915432	81491598	14387687
14698765	76159876	38769872
71687692	40536987	65437652
14187165	68769824	87691874

(36)	(37)	(38)
24809	34875926	4
45	4875926	24
6265	875926	879
78	75926	4262
789432	5926	58715
47	75926	246860
1663	875926	1967748
29	4875926	46846345
4934781	34875926	568596385

(39)	(40)	(41)
5682436	14868321	9876543210
185	57	987654321
5682436	1208	98765432
9185	56110	9876543
5682436	8920965	987654
79185	2	98765
5682436	10	9876
85	487	987
5682436	2675	98
479185	34158	9

42. Add together 647,500,698 ; 4,130 ; 87,592 ; 401 ; 3,698,762 and 1,097.

43. What is the sum of 609 ; 2,413 ; 85 ; 10,001 ; 6,849,100,609 and 5,000 ?

44. Find the amount of 24,010 ; 8,956 ; 704,964 ; 8,007,659 and 10.

45. Add 24 ; 687 ; 2,419 ; 18,765 ; 769,827 ; 1,208,792 and 29 together.

46. Add together 246,987,065,423 ; 809,654 ; 298,762,423 ; 50,097 and 5.

47. What is the amount of 24,009 ; 809,692,483 ; 69,821 ; 34 and 160 ?

48. What is the amount of eight hundred ; seventy thou sand six hundred ; two millions ninety thousand and eight ; twenty millions two hundred thousand and thirty-five ; thirty millions eight thousand and two ; and seven hundred fifteen ?

49. Find the sum of two trillions four hundred billions nine hundred forty-five thousand six hundred twenty-three ; four hundred billions twenty millions six thousand and seventy ; eighty-four thousand ; and six hundred eighty-four.

50. Add together seven hundred sixty ; eighty-nine thousand forty-nine ; two hundred fifty-eight millions six hundred twenty-four ; seven thousand eight hundred twenty-two ; and sixteen.

51. Add together two trillions eight millions six hundred thousand and nine; eight trillions four hundred seventy thousand two hundred forty-two; nine hundred forty-nine millions six hundred forty-seven; and seven hundred eighty-five.

52. What is the sum of seven hundred fifty-two; nine thousand three hundred; and eight millions?

53. Find the sum of twenty-nine millions eight hundred forty-eight thousand six hundred ninety-five; seven thousand two hundred eighty; and six hundred ninety-five.

54. Add together two hundred ninety-seven; eight thousand sixty-four; and seven hundred millions.

55. Add together sixteen; two hundred forty-five; seventy-five thousand nine hundred; and twenty.

56. Add together forty-nine thousand two hundred twenty; eighty-five thousand six hundred and seven; ninety-nine millions; and forty-two.

57. What is the sum of sixty-nine; eighty-four thousand seventy-nine; two hundred forty-eight thousand; and fifty-nine?

58. Add together seventy-nine thousand two hundred forty; fifty-eight millions sixty-five thousand; one hundred fifty; and twenty-seven.

59. What is the amount of seven hundred twenty-five; sixty-eight thousand; ninety-five millions; and five?

60. What is the sum of twenty-eight thousand; seven thousand sixty; two hundred forty; fifty-seven; and nine?

61. What is the sum of two millions five hundred eighty-five thousand seven hundred and ninety; eight hundred ninety-seven thousand; and seven thousand two hundred ten?

62. Add together six; twenty-five; four hundred fifteen, seven thousand six hundred twenty; ninety-four thousand one hundred and five; and six millions.

63. Find the amount of two millions and six; eight thousand; and two hundred.

64 Add together nine hundred fifteen; twenty-five millions two hundred; eight hundred sixty-five thousand; forty-five thousand seventy-five; two hundred sixty eight; and one thousand.

65 Find the sum of ninety-nine millions six hundred thousand; eight millions four hundred and ten; seven hundred four thousand and seventy; sixty-five thousand two hundred; five thousand and one; four hundred seventeen; thirty-six; and two.

SUBTRACTION.

SUBTRACTION is the process of finding the difference between two numbers.

The greater or upper number in Subtraction is called the Minuend ; the less or lower number, the Subtrahend ; and the answer is called the Difference or Remainder.

(1)	(2)
From 65437982	4756781932
Take 42136321	1235441512
A. 23301661	

(3)	(4)
87916543267	598436987461
19876274382	479658279147

(5)	(6)
2469827654	87291468782
2387696273	49627319872

(7)	(8)
38696754794	135879654368
14982965437	118296547187

(9)	(10)
46987692436	478692714687
31498769765	143789671582

(11)	(12)
200936874415	2000870069871
199876921562	1000741587692

(13)	(14)
1020304050607	500000649876
189060509081	200000878694

(15)	(16)
4287690000624	100000000000
1598746982009	99999999999

(17)	(18)
100030005000600	45671000000
80002000900	14960000001

(19)	(20)
230060087690008	1000006000081
230040091006204	48764001906

(21)	(22)
976847620498	300400500600
769259768149	100900800911

(23)	(24)
100006800009	100000870047
10009176872	1987000928

25. From 10,000,000 take 9.

26. From 7,062,481 take 2,400.

27. From 2,048,197 take 2,047,822.

28. From 16,006 take 6,006.

29. Find the difference between 180,009,621,001 and 10,002.

30. What is the difference between 100,809,658,762 and 218,769?

31. What is the difference between 87,699,821,976 and 27,698,741,468?

32. Subtract 30,000,687 from 42,876,512.

33. Subtract 2,681 from 5,000.

34. From 24,000,089 take 14,000,071.

35. From 5,090,875 take 2,065,784.

36. From 16,000,000 take 59.

37. From 204,080,249 take 81,698.

38. From 121,000,604 take 980.

39. From 4,187,692 take 2,849,126.

40. Find the difference between 100,000,849,101 and 249,687,614,185.

41. Find the difference between 240,008 and 2,918,764,-198,001.

42. Find the difference between 80,491,006 and 70,456,-
829.

43. Find the difference between 15,871,692,043 and
10,000,009.

44. From 10,004,905 take 60,495.

45. From 203,004,000 take 2,001,000.

46. From seven millions and eighty-nine, take eighty
thousand and five.

47. From eight hundred nineteen millions, take eighty-
seven thousand.

48. From two hundred billions seven hundred sixty-five
millions eight hundred twenty-seven, take one billion
thirty-four millions three thousand.

49. From fifty-five thousand, take four hundred and two.

50. From one hundred billions one hundred thousand one
hundred, take one thousand and ninety-nine.

51. From twelve billions nineteen thousand and two, take
twelve billions eighteen thousand and nine.

52. From twenty-five trillions, take twenty billions six
hundred eighty-nine.

53. From twenty-nine billions two hundred forty-nine
thousand six hundred twenty-four, take nineteen billions two hundred forty-nine thousand five hundred
eighty-five.

54. From two hundred one millions six hundred twenty-five, take nine hundred ninety-nine.

55. From one hundred one millions nine thousand, take ninety-nine.

56. From two thousand twenty, take one hundred.

57. From eighty-five billions six hundred forty take nine hundred eighty-five thousand and four.

58. From seven hundred sixty-nine millions two hundred seventy-five thousand, take eighty-seven thousand.

59. From two hundred thousand, take ninety-nine thousand.

60. From one hundred take eleven.

61. From seven hundred four millions eighty-one thousand two hundred and five, take ninety-nine millions and six.

62. From one hundred eleven millions six hundred fifty-seven, take ten millions six hundred fifty-six.

63. From twelve millions, take twelve thousand.

64. From one hundred eighty-five billions one hundred eighty-five, take one hundred eighty-five thousand.

65. From ten thousand, take nineteen.

66. From twenty-five millions, take four hundred thousand and sixteen.

67. From nineteen millions fourteen thousand two hundred fifteen, take seventeen millions and forty-five. –

MULTIPLICATION.

MULTIPLICATION is the process of taking one of two numbers as many times as there are units in the other.

The number to be multiplied is called the Multiplicand; the number by which we multiply, the Multiplier; and the answer is called the Product.

The multiplicand and multiplier are called the Factors of the product.

	(1)		(2)
Multiply	6487521		984763571
By	2		3
A.	12975042		

3. Multiply 594,876,952 by 4.

4. Multiply 437,628,741 by 5.

5. Multiply 159,870,623 by 6.

6. Multiply 287,432,718 by 7.

7. Multiply 146,527,384 by 8.

8. Multiply 376,928,945 by 9.

9. Multiply 628,496,587 by 11.

10. Multiply 984,657,234 by 12.

(11)	(12)
854638721	249876354
10	200
8546387210	49975270800

(13)	(14)
487698762	76943762
34	506
1950795048	461662572
1463096286	384718810
16581757908	38933543572

(15)	(16)
687954	23005
70008	1000900
5503632	20704500
4815678	23005
48162283632	23025704500

17. Find the product of 240,957 and 2400.

18. Find the product of 89,076,541 and 130.

19. Find the product of 14,268,759 and 1090.

20. Multiply 8,465,792 by 14.

21. Multiply 290048,756 by 160.

22. Multiply 159,467,487 by 38.

23. Multiply 300,897,692 by 27.

24. Multiply 473,287,641 by 18.

25. Multiply 3,040,806,001 by 59.

26. Multiply 15,469,820 by 97.

27. Multiply 80,592,417 by 86.

28. Multiply 14,658,769 by 345.

29. Find the product of 67,592,416 and 605.

30. Find the product of 18,700 and 476,908,732.

31. Find the product of 1,369,008,430 and 20,057.

32. Find the product of 2,090,300 and 69,010,000.

33. Find the product of 2,845,000 and 29,045,600.

34. Multiply 80,469,827 by 123.

35. Multiply 19,876,431 by 678.

36. Multiply 4,165,927 by 999.

37. Multiply 123,456,789 by 2080.

38. Multiply 2,046,827 by 419.

39. Multiply 2,160,008 by 367

40. Multiply 394,996 by 455.

41. Multiply 1,876,977 by 636.

42. Multiply 424,484 by 8828.

43. Multiply 6,629,666 by 9222.

44. Multiply 88,187,888 by 6789.

45. Multiply 55,552,222 by 4567.

46. Multiply 67,892 by 13,458.

47. What is the product of 6,948,756,297 and 49,800?

48. What is the product of 8,946,537,128 and 24,715?

49. Multiply twenty-four thousand eight hundred, by nine hundred and sixty.

50. Multiply eighteen millions and ten, by six hundred forty-eight thousand.

51. Multiply seventy-five thousand, by seventy-five thou· sand.

52. Multiply one thousand eight hundred and ninety, by ninety-four.

53. Multiply two hundred fifty-eight, and seven hundred together.

54. Multiply eight thousand ninety-six, and seventy thousand together.

55. Multiply fifty-nine billions, by eight hundred and sixteen.

56. Multiply twenty-two thousand and sixty-five, by twelve billions.

57. Multiply ninety-nine trillions, by eight.

58. Multiply sixty-seven billions eighty-four thousand, by twenty-five.

59. Multiply nine hundred eighty-four, by six hundred fifty-seven.

60. Multiply one thousand and seven, by five thousand and ninety.

61. Multiply sixty-nine billions and six, by two hundred and forty-nine.

62. Multiply eight hundred forty-six millions, by five hundred.

DIVISION.

DIVISION is the process of finding how many times one number is contained in another.

The number to be divided is called the Dividend; the number by which we divide, the Divisor; and the answer is called the Quotient.

If the dividend does not contain the divisor an even number of times that part which is left is called the Remainder. This must always be less than the divisor.

SHORT DIVISION.

1. Divide 68462048 by 2.

$$2 \,)\, 68462048$$

Ans.   34231024

(2)

$$3 \,)\, 693063$$

(3)

$$2 \,)\, 128469247$$

Ans.   64234623—1 rem

4. Divide 2,587,695 by 3.

5. Divide 23,468,049 by 4.

6. Divide 687,649,325 by 5.

7. Divide 847,926,592 by 6.

8. Divide 147,968,563 by 7.

9. Divide 4,825,926,374 by 8.

10. Divide 7,648,237,654 by 9.

11. Divide 1,239,018 by 11.

12. Divide 21,025,388 by 12.
13. Divide 680,195,472 by 4.
14. Divide 376,584,751 by 8.
15. Divide 18,927,054 by 6.
16. Divide 8,476,389,217 by 2.
17. Divide 3,172,987,643 by 7.
18. Divide 49,811,956 by 12.
19. Divide 654,237,846 by 3.
20. Divide 215,609,840 by 5.
21. Divide 2,100,967 by 11.
22. Divide 143,768,547 by 9.

(23)	(24)
1|0) 2468975|2	2|0) 1876854|0
2468975—2 rem.	93827

25. Divide 39,876,540 by 30.
26. Divide 872,694,180 by 400.
27. Divide 187,643,587 by 500.
28. Divide 24,976,400 by 6000.
29. Divide 19,687,460 by 700.
30. Divide 21,596,329 by 80.
31. Divide 912,368,200 by 9,000.
32. Divide 8,469,420 by 110. .
33. Divide 48,108,240 by 120.
34. Divide 24,695,700 by 12,000,000.
35. Divide 881,056,240 by 1100.

LONG DIVISION.

36. Divide 146928652 by 13.

```
        13) 146928652 (11302204
            13
            ──
            16
            13
            ──
            39
            39
            ──
            28
            26
            ──
            26
            26
            ──
            52
            52
            ──
            00
```

37. Divide 159,476,328 by 14.

38. Divide 207,598,675 by 15.

39. Divide 198,263,594 by 16.

40. Divide 217,624,380 by 17.

41. Divide 254,963,872 by 18.

42. Divide 232,746,358 by 19.

43. Divide 3,654,876,394 by 25.

44. Divide 91,236,187,432 by 48.

45. Divide 6,927,387,420 by 36.

46. Divide 10,982,763,587 by 53.

47. Divide 87,492,659,218 by 64.

48. Divide 14,023,784,600 by 72.

49. Divide 6,432,789,741 by 89.

50. Divide 1,127,436,287 by 91.

51. Divide 824,676,324 by 587.

52. Divide 207,360 by 144.

53. Divide 2,576,904 by 258.

54. Divide 769,422,078 by 7245.

55. Divide 172,865 by 1728.

56. Divide 2,646,209 by 25,462.

57. Divide 322,400 by 248.

58. Divide 320,400 by 36.

59. Divide 260,050 by 25.

60. Divide 24,681,234 by 1234.

61. Divide 101,496,875 by 11,982.

62. Divide 2,379,874,682 by 36,845.

63. Divide 158,673,982,760 by 149,824.

64. Divide 15,184,340 by 505.

65. Divide 10,201 by 101.

66. Divide 8,467,923,846 by 319,425.

67. Divide 199,677,792 by 974.

68. Divide 987,634,092 by 987.

69. Divide 108,060,704 by 10,406.

70. Divide 372,415,965 by 2,483,712.

71. Divide 10,199,932 by 849,161.

2. Divide 483,765 by 294,186.

73. Divide 382,941,962 by 124.
74. Divide 411,690,480 by 137.
75. Divide 182,674,387 by 1,384,692.
76. Divide 874,639,746,358 by 2,345.
77. Divide three hundred ninety millions, by thirteen.
78. Divide seven hundred eighteen thousand six hundred by nineteen.
79. Divide four thousand seven hundred ninety-two, by twenty-seven.
80. Divide sixty-eight millions and seven, by five thousand.
81. Divide two hundred fifty-four thousand seven hundred, by twenty-four.
82. Divide two thousand eight hundred eighty, by twelve.
83. Divide six hundred forty billions and sixty-four, by sixteen.
84. Divide eighteen millions sixteen thousand, by two hundred and forty.
85. Divide three hundred twenty-eight thousand six hundred seventy-two, by seventeen thousand and one.
86. Divide twenty-four millions and sixteen, by twenty thousand.
87. Divide two hundred fourteen thousand, by one hundred and ten.
88. Divide twenty-nine trillions, by fourteen.
89. Divide five hundred forty millions two hundred ten thousand and six, by one hundred twenty thousand

MENTAL EXERCISES.*

4. 1. If you had five cherries, and Mary should give you four more : how many would you have then ?

2. Edward found six eggs in one nest, and four in another : how many eggs did he find ?

3. If you saw four butterflies fluttering about one flower, and three about another, how many butterflies would you see ?

4. Ralph let his sister Grace put her books into his bag. He had six of his own, and she put in three. How many books did he have to carry ?

5. Charles had five cents, and his uncle gave him six more : how many had he then ?

6. There were seven cows in one field, and five in another : how many in both ?

7. Jane had ten chickens, and her mother gave her three more : how many had she then ?

8. Mary had four peaches, George had six, and little Bessie had two : how many had they all together ?

---

* It is recommended that the pupils be not required to study these lessons before being called to the class. Let the teacher use the recitation as a means of testing and guiding the computing powers of the pupils, and of inducting their minds into the simplest processes of reasoning.

9. Daniel, Michael and William went fishing. Daniel caught eight fishes, Michael caught six, and William, three. How many fishes were caught in all?

10. Bertha had five dolls, Martha had three, and Betsey had two. They set them all in a row against the wall, and counted them. How many did they count?

11. Three little boys went nutting. John got one pint, Peter seven pints, and Herbert four pints. How many in all?

12. If your father sent you for seven pounds of flour, four pounds of sugar, and one pound of tea, how many pounds would you have to carry home?

13. There was a great tornado, and it blew down five houses on one street, seven on another, and four on another: how many houses did it blow down?

14. Harry took care of six sheep, two lambs, eight chickens, and three turkeys: how many animals did he take care of?

15. If you had two pockets, and had six pears in each pocket and four in your hat, how many pears would you have?

16. A cat saw a rat and started for it. Then three more cats ran up to see what *they* could find A dog saw them and began to bark furiously, and two more dogs came round the corner to see what the fuss was about. Dogs, cats, and rat, how many were there?

17. Four little boys played marbles together. Their names were Robert and Morris and Jamie and Willie. Robert had seven marbles, Morris had three, Jamie had ten, and Willie had eight. When they stopped playing they said, Let us put all our marbles together in a box. They did so. How many marbles were there in the box?

B. 1. If there were seven cherries on a stem, and you ate three of them, how many would be left?

2. If you caught ten crabs, and five of them fell back into the water, how many would you have to carry home?

3. If six little boys were sitting on a fence, and two fell off, how many would be left on the fence?

4. Little Lulu's mother gave her twelve peanuts, and she gave her sister Myra six of them: how many did she have left?

5. Charles put fourteen nuts into his pocket, but there was a hole in it, and all but five dropped out: how many did he lose?

6. Nine caterpillars were crawling on the ground. A boy threw a stone and killed three of them. How many were left alive?

7. James gave a blind man fifteen pennies, and he spent seven of them for bread: how many did he have left?

8. Willie had ten marbles, and John had three : how many more had Willie than John ?

9. Sallie had six apples, and gave Etta all but one: how many did she give her ?

10. There were seventeen trees in one orchard, and nine in another : how many more in the first than in the second ?

11. "How dark it is !" said little George ; "there are no stars to be seen." "O yes ! there are," said his brother Dick ; "I can see six." "And I," said their cousin William, "can see ten." How many more did William see than Dick ?

12. There were eight bunches of grapes in a fruit-dish, and Emma's mother told her to put three of them on a plate, and take them to her aunt Jane : how many were left in the fruit-dish ?

13. Mary went to the store to buy some peaches for her mother. Her mother gave her sixteen cents, and the peaches cost nine cents. How many cents ought she to have taken back ?

14. Mary did her errand so well that her mother sent her again, to buy some cheese. She took the money which she brought back the first time, and the cheese cost just seven cents. How much did she have for her mother this time ?

15. If one window had sixteen panes of glass in it, and another had eight, how many more would the first window have than the second ?

16. Ann had twelve books in her desk, and Florence had five: how many more had Ann than Florence?

17. It took Peter fifteen minutes to read a page in his book, but Isaac could read the same page in six minutes: how much longer did it take Peter than Isaac?

C. 1. If I had three rabbits, and you had three times as many, how many would you have?

2. Nora told Gertrude that she had six roses on her rose-bush, but Gertrude said that there were twice as many on hers: how many roses on Gertrude's rose-bush?

3. Henry saw a man selling peanuts. He asked him how much they were. The man said, "Six cents a pint." "Then," said Henry, "I will take four pints." How many cents did he give the man?

4. If a squirrel eats eight nuts in one day, how many nuts will it eat in eight days?

5. If a little boy eats five pieces of bread every day, how many pieces will he eat in a week?

6. Little Eddie came home crying because he had lost four chestnuts. "Don't cry," said his big brother Jack; "I will give you four times as many as you have lost." How many did he give him?

7. Fanny had eight books, and Sarah had five times as many: how many books had Sarah?

8. There are four classes in the school, and ten boys in each class: how many boys in the school?

9. There was a heavy snow-storm to-day, and only six girls were at school; but there are seven times as many when it is pleasant: how many are there when it is pleasant?

10. Six benches in a row, and nine ladies and gentlemen on each bench: how many on all of them?

11. Little Rose's father gave her a box of houses, and one day she made a big town on the floor. There were five streets in her town, and on every street were ten houses. How many houses were there in the town?

12. Joseph climbed the cherry-tree and plucked seven clusters of cherries. There were five cherries in each cluster: how many in all of them?

13. Michael bought six loaves of bread, costing eight cents a loaf: how many cents did he pay?

14. John's father said to him, "I will give you on Christmas three times as much money as you will earn before that time." He earned nine cents: how much did his father give him?

15. Ella's school-house was two miles off. She walked to and from it every day, five days in the week. How many miles did she walk every week?

16. There was a dining-room containing eight tables, and six persons were at each table : how many persons in all ?

D. 1. Mary brought home eight apples to her four little sisters : how many did she give to each ?

2. If you had to give one cent for one apple, how many apples could you buy for four cents ?

3. If you have six bananas, and eat two every day, how many days will they last ?

4. In an orchard there were twelve trees, standing in rows, and in each row there were four trees : how many rows were there ?

5. Sarah boiled ten eggs for breakfast. There were five persons in the family : how many eggs did each one have ?

6. If your father gave you four pennies every week, how many weeks would it take you to get twenty pennies ?

7. If it costs five cents to ride in the horse-car, how many times can you ride for twenty-five cents ?

8. If Ansel had twenty-seven marbles, and he could put nine in a box, how many boxes would it take to hold them all ?

9. If you can make three whistles in a day, how many days will it take you to make twenty-four ?

10. If a man can walk four miles an hour, in how many hours can he walk twenty miles ?

11. Clementine had ten dolls, and she put them in two rows : how many dolls were there in each row?

12. How many kites, at four cents apiece, can you buy for twelve cents?

13. If one plate will hold just six apples, how many plates must you have for eighteen apples?

14. Dennis had sixteen carriage-wheels: how many carriages would they supply, if each carriage needed four wheels?

15. If four boys can ride in one cart, how many carts will be necessary if twelve boys want to ride?

16. There were twelve marbles and three boys. If the marbles were divided equally among the boys, how many would each boy have?

E. 1. How many are two and four? Twelve and four? Twenty-two and four? Thirty-two and four? Forty-two and four? Fifty-two and four? Sixty-two and four? Seventy-two and four? Eighty-two and four? Ninety-two and four?

2. How many are three and six? Thirteen and six? Twenty-three and six? Thirty-three and six? Forty-three and six? Fifty-three and six? Sixty-three and six? Seventy-three and six? Eighty-three and six? Ninety-three and six?

3. How many are five and two? Fifteen and two?
   Twenty-five and two? Thirty-five and two?
   Forty-five and two ? Fifty-five and two ? Sixty-
   five and two? Seventy-five and two? Eighty-
   five and two ? Ninety-five and two?

4. How many are two and ten ? Twelve and ten?
   Twenty-two and ten? Thirty-two and ten?
   Forty-two and ten ? Fifty-two and ten ? Sixty-
   two and ten? Seventy-two and ten? Eighty-
   two and ten ? Ninety-two and ten?

5. How many are five and five? Fifteen and five?
   Twenty-five and five? Thirty-five and five?
   Forty-five and five ? Fifty-five and five ? Sixty-
   five and five? Seventy-five and five ? Eighty-
   five and five ? Ninety-five and five?

6. How many are four and seven? Fourteen and
   seven? Twenty-four and seven? Thirty-four
   and seven? Forty-four and seven? Fifty-four
   and seven? Sixty-four and seven? Seventy-
   four and seven ? Eighty-four and seven ? Ninety-
   four and seven ?

7. How many are six and four? Sixteen and four?
   Twenty-six and four? Thirty-six and four?
   Forty-six and four ? Fifty-six and four? Sixty-
   six and four? Seventy-six and four? Eighty-
   six and four? Ninety-six and four?

8. How many are .ten and five ? Twenty and five ?
   Thirty and five? Forty and five? Fifty and

five? Sixty and five? Seventy and five? Eighty and five? Ninety and five?

9. How many are three and eight? Thirteen and eight? Twenty-three and eight? Thirty-three and eight? Forty-three and eight? Fifty-three and eight? Sixty-three and eight? Seventy-three and eight? Eighty-three and eight? Ninety-three and eight?

10. How many are nine and five? Nineteen and five? Twenty-nine and five? Thirty-nine and five? Forty-nine and five? Fifty-nine and five? Sixty-nine and five? Seventy-nine and five? Eighty-nine and five? Ninety-nine and five?

11. How many are four and nine? Fourteen and nine? Twenty-four and nine? Thirty-four and nine? Forty-four and nine? Fifty-four and nine? Sixty-four and nine? Seventy-four and nine? Eighty-four and nine? Ninety-four and nine?

12. How many are six and ten? Sixteen and ten? Twenty-six and ten? Thirty-six and ten? Forty-six and ten? Fifty-six and ten? Sixty-six and ten? Seventy-six and ten? Eighty-six and ten? Ninety-six and ten?

13. How many are seven and eight? Seventeen and eight? Twenty-seven and eight? Thirty-seven and eight? Forty-seven and eight? Fifty-seven and eight? Sixty-seven and eight? Seventy-

seven and eight ? Eighty-seven and eight ?
Ninety-seven and eight ?

14. How many are six and eight ? Sixteen and eight ?
Twenty-six and eight ? Thirty-six and eight ?
Forty-six and eight ? Fifty-six and eight ? Sixty-
six and eight ? Seventy-six and eight ? Eighty-
six and eight ? Ninety-six and eight ?

15. How many are nine and nine ? Nineteen and nine ?
Twenty-nine and nine ? Thirty-nine and nine ?
Forty-nine and nine ? Fifty-nine and nine ?
Sixty-nine and nine ? Seventy-nine and nine ?
Eighty-nine and nine ? Ninety-nine and nine ?

16. How many are nine and ten ? Nineteen and ten ?
Twenty-nine and ten ? Thirty-nine and ten ?
Forty-nine and ten ? Fifty-nine and ten ? Sixty-
nine and ten ? Seventy-nine and ten ? Eighty-
nine and ten ? Ninety-nine and ten ?

F. 1. Maggie had ten little turkeys. Four of them
died, and three ran away. How many did
Maggie have left ?

2. Ten little girls were playing tag. Four wore blue
dresses ; two, red dresses, and all the rest wore
white dresses. How many were in white ?

3. James is seven years old, and his sister Clementine
is three years younger : how old is Clementine ?

4. In four years how old will James be ?

5. How old will his sister be ? How much older will James be than Clementine then?

6. Twenty boys were walking in a procession. They met a man playing a hand-organ, and four stopped to listen to the music. Then somebody cried *fire*, and eight ran to see where the fire was. How many were left walking in the procession ?

7. If I have four horses, and I buy six more of one man and three of another, how many will I have then ?

8. If now I sell six, how many will I have left ?

9. In a certain school there were eleven boys and eight girls : how many pupils were there?

10. One day four boys and two girls were absent: how many were at school ?

11. How many boys? How many girls?

12. If you have twelve turkeys, and sell four and give away three, how many will you have left?

13. Maria wrote seventeen lines in her copy-book. Her teacher told her that ten were well written, but that the others were so badly done that she must write them over again. How many did she have to write a second time?

14. How many lines did she write all together?

15. Benjamin found eight ripe apples under one tree, six under another, and four under another : how many did he find under the three trees?

13. He gave nine to his brother and kept the rest himself: how many did he keep?

17. In a school there are eighteen boys and fourteen girls: how many more boys than girls?

18. If six boys and two girls stayed at home, how many more boys than girls then?

19. Harry, Russell, and Percy went to the orchard to gather peaches. Harry gathered sixteen; Russell, ten; and Percy, eight. On the way home Percy upset his basket, and a horse and wagon ran over the peaches and spoiled them all; so Harry gave him five of his, and Russell gave him three: how many had he then?

20. How many did Harry have left? How many did Russell have left?

21. A procession, consisting of six hundred men, met a procession of four hundred, and they concluded all to march together: how many men in the new procession?

22. Afterwards five hundred of the men went home: how many were left?

23. A boy had six marbles, and one boy gave him five, another gave him three, and another enough to make his number twenty. How many did the last boy give him?

G. 1. Eight and four, less three, are how many?

2. Six and five, less four, are how many?

3. Seven and nine, less five, are how many?

4. Three and seven, less six, are how many?

5. Nine and five, less seven, are how many?

6. Ten and seven, less eight, are how many?

7. Twelve and three, less nine, are how many?

8. Eleven and eight, less ten, are how many?

9. Three, and five, and nine, less seven, are how many!

10. Four, and six, and ten, less eleven, are how many!

11. Fourteen, and six, and eight, less eight, are how many?

12. Five, and five, and eight, less ten, are how many?

13. Eight, and nine, and ten, less eight, are how many?

14. Seven, and seven, and eight, and nine, less seven, are how many?

15. Fifteen, and five, and seven, and six, less ten, are how many?

16. Twelve, and nine, and six, and five, less six, are how many?

17. Eighteen, and twelve, and nine, and ten, less nine, are how many?

18. Twenty, and seven, and nine, and eight, less five, are how many?

19. Fifteen, and ten, and eight, and six, and nine, less two, are how many?

20. Twenty-five, and nine, and six, and seven, and eight, less ten, are how many?

21. Nineteen, and eight, and seven, and four, and six, less five, are how many?

22. Thirty, and eight, and five, and ten, and seven, less ten, are how many?

23. Eighteen, and nine, and five, and four, less nine, are how many?

24. Twenty-nine, and ten, and eight, and six, and nine, less three, are how many?

25. Thirty, and twelve, and six, and ten, and two, less five, are how many?

26. Thirty-six, and seven, and eight, and six, less seven, are how many?

H. 1. Samuel went to the fruit-store with thirty cents. He bought five pears at two cents apiece, and three oranges at four cents apiece: how much did he pay for them all?

2. He took the change home to his father, who said it was correct: how much was it?

3. The father then divided the change equally between Samuel and his brother: how much did each have?

4. Albert and Caroline went blackberrying. Albert picked four quarts of blackberries, and Caroline picked three times as many wanting two quarts: how many did Caroline pick?

5. A woman had five pounds of butter to sell. She sold four pounds for twenty cents a pound, and one pound for twenty-five cents: how much money did she receive?

6. Henry worked five days for three dollars a day, and John worked three days for two dollars a day: how much more did Henry earn than John?

7. Charlotte and her cousin Jimmie went to take a walk. Charlotte showed Jimmie four lambs. Jimmie told her he would show her three times as many cows: how many cows did he show her?

8. If you should count the cows and the lambs together, how many would there be?

9. After they had seen the cows they went to look at some birds. There were thirty-two birds, four in a cage: how many cages were there?

10. Augustus, Willie, and Robert had each of them four cents. If they put all their money together, how many cakes could they buy with it, if cakes were two cents apiece?

11. A little girl had a party, and invited eighteen dolls. She said that three dolls must sit together on a chair: how many chairs did she prepare for them?

12. The day turned out to be rainy, and nine dolls stayed at home: how many dolls came? How many chairs did they need?

13. She gave them two plums apiece, but the dolls would not eat them, so she and her sister ate them up themselves : how many did they eat?

14. If they ate an equal number, how many did each eat?

15. I bought a barrel of apples for two dollars, and a barrel of cider for four dollars. I sold the apples for three dollars, and the cider for six dollars · how much did I make on my purchases?

16. Oliver answered eight questions in history and Roland answered three times as many : how many did Roland answer? How many more did Roland answer than Oliver? How many did both together answer ?

17. Three boys talking of their marbles, one said he had nine ; another said, he had twice as many ; and the other said, he had as many as both of them put together : how many had the last ?

18. Willie had fourteen cents ; Adelaide, ten ; Almira, six ; Lulu, five ; and Ethel, five : how many had they in all ?

19. Each of them put four cents into a box : how many cents were in the box? How many cents were left ?

20. How many cents did each one have left ?

21. They gave the box to Adelaide and told her to divide the money equally among four poor children : how much did each child receive ?

22. James bought of a baker three rolls at a cent apiece, two bunns at two cents apiece, and a piece of gingerbread for six cents. He gave the baker twenty-five cents: how much change should he receive?

23. Ada bought three ounces of candy at five cents an ounce, four cakes at three cents apiece, and some mottoes for ten cents. When she came to pay for her purchases, she found that she had but thirty cents: how many more cents did she need?

24. Fred has fourteen things in his pocket, and his sister has ten in hers. If Fred gives his sister three of his things and the sister gives him four of hers, how many will be in each of the pockets then?

25. Hermann had forty cents. He spent it for apples at two cents apiece, and divided the apples equally between his three brothers and himself: how many did each have?

26. A man bought twenty oranges at four cents apiece, and paid for them in ten-cent pieces: how many ten-cent pieces did he pay?

MISCELLANEOUS QUESTIONS.

1. How many days in the week? *Ans.* Seven.
2. Name them. *Ans.* Sunday, Monday, Tuesday, Wednesday, Thursday, Friday, Saturday.
3. How many months in the year? *Ans.* Twelve.
4. Name them. *Ans.* January, February, March, April, May, June, July, August, September, October, November, December.
5. How many seasons in the year? *Ans.* Four.
6. Name them. *Ans.* Spring, Summer, Autumn, Winter.
7. How many days in each month?
   Ans. Thirty days hath September,
   April, June, and November.
   ᴀ ll the rest have thirty-one,
   Save February, which alone
   Hath twenty-eight; and this, in fine,
   Each leap-year hath twenty-nine.
8. How many single things make a dozen?
   Ans. Twelve
9. How many dozen make a gross? *Ans.* Twelve
10. How many single things make a score?
    Ans. Twenty,
11. How many sheets of paper make a quire?
    Ans. Twenty-four
12. How many quires make a ream? *Ans.* Twenty
13. How many feet make a fathom? *Ans.* Six.
14. How many inches make a hand? *Ans.* Four.

· CURRENCY, AND WEIGHTS AND MEASURES.

UNITED STATES MONEY.

10 mills	make	1 cent.
10 cents	"	1 dime.
10 dimes	"	1 dollar.
10 dollars	"	1 eagle.

ENGLISH MONEY.

4 farthings	make	1 penny.
12 pence	"	1 shilling.
20 shillings	"	1 pound or sovereign.

TROY WEIGHT.

24 grains	make	1 pennyweight.
20 pennyweights	"	1 ounce.
12 ounces	"	1 pound.

AVOIRDUPOIS WEIGHT.

16 ounces	make	1 pound.
100 pounds	"	1 hundred-weight.
20 hundred-weight, = 2000 pounds, make 1 ton.		

APOTHECARIES' WEIGHT.

20 grains	make	1 scruple.
3 scruples	"	1 dram.
8 drams	"	1 ounce.
12 ounces	"	1 pound.

LIQUID MEASURE.

4 gills	make	1 pint.
2 pints	"	1 quart.
4 quarts	"	1 gallon.
31½ gallons	"	1 barrel.
2 barrels, or 63 gallons,	"	1 hogshead.
2 hogsheads	"	1 pipe or butt.
2 pipes, or 4 hogsheads,	"	1 tun.

DRY MEASURE.

2 pints	make	1 quart.
8 quarts	"	1 peck.
4 pecks	"	1 bushel.

LONG MEASURE.

12 inches	make	1 foot.
3 feet	"	1 yard.
5½ yards, or 16½ feet,	"	1 rod.
320 rods	"	1 statute mile

SQUARE MEASURE.

144 square inches	make	1 square foot.
9 square feet	"	1 square
30¼ square yards	"	1 sq. keg.
160 square rods	"	1 barrel.
640 acres	"	1 barrel.
	"	1 cask.

SURVEYOR'S LONG MEASURE.

7.92 inches	make	1 link.
25 links	"	1 rod.
4 rods, or 66 feet,	"	1 chain.
80 chains	"	1 mile.

SURVEYOR'S SQUARE MEASURE.

625 square links	make	1 pole.
16 poles	"	1 square chain.
10 square chains	"	1 acre.
640 acres	"	1 square mile.
36 square miles(6 miles square)	"	1 township.

CUBIC MEASURE.

1728 cubic inches	make	1 cubic foot.
27 cubic feet	"	1 cubic yard.
40 cubic feet of round timber	make	1 ton or load
50 cubic feet of hewn timber	"	1 ton or load
16 cubic feet	make	1 cord foot.
8 cord feet	"	1 cord of wood.
128 cubic feet	"	1 cord of wood.
24¾ cubic feet	"	1 perch of stone or masonry.

TIME MEASURE.

20 grs.		
3 scruples	make	1 minute.
8 drams	"	1 hour.
12 ounces	"	1 day.

7 days	make	1 week.
365 days	"	1 common year.
366 days	"	1 leap year.
12 calendar months	"	1 year.
100 years	"	1 century.

CIRCULAR MEASURE.

60 seconds	make	1 minute.
60 minutes	"	1 degree.
30 degrees	"	1 sign.
12 signs, or 360 degrees,	"	1 circle.

LONGITUDE AND TIME.

15 degrees of longitude	equal	1 hour of time.
15 minutes of longitude	"	1 minute of time.
15 seconds of longitude	"	1 second of time.

MISCELLANEOUS TABLES.

COUNTING.

12 units, or things,	make	1 dozen.
12 dozen	"	1 gross.
12 gross	"	1 great gross.
20 units	"	1 score.

100 pounds of nails	make	1 keg.
196 pounds of flour	"	1 barrel.
200 pounds of pork or beef	"	1 barrel.
240 pounds of lime	"	1 cask.

24 sheets of paper	make	1 quire.
20 quires "	"	1 ream.
2 reams "	"	1 bundle.
5 bundles "	"	1 bale.

A sheet folded in-

2 leaves	is called	a folio, and 1 sheet of paper makes 4 pages.
4 "	"	a quarto, or 4to, and 1 sheet of paper makes 8 pages.
8 "	"	an octavo, or 8vo, and 1 sheet of paper makes 16 pages.
12 "	"	a duodecimo, or 12mo, and 1 sheet of paper makes 24 pages.
16 "	"	a 16mo, and 1 sheet of paper makes 32 pages.
18 "	"	an 18mo, and 1 sheet of paper makes 36 pages.

ANSWERS TO SLATE EXERCISES.

ADDITION.

Example.		Example.	
3	1,829,687,262	35	425,176,635
4	131,385,964	36	5,757,149
5	12,564,256	37	81,413,334
6	175,363,302	38	617,741,222
7	116,624,940	39	28,980,005
8	16,437,386	40	23,883,993
9	1,856,909,063	41	10,973,936,895
10	82,000,979	42	651,292,680
11	1,382,676	43	6,849,118,717
12	205,025,556	44	8,745,599
13	201,162,984	45	2,000,543
14	226,436,283	46	247,286,687,602
15	3,068,809,613	47	809,786,507
16	210,048,229	48	52,370,164
17	3,188,594,280	49	2,800,021,036,377
18	2,765,999,216	50	258,098,271
19	14,011,240	51	10,000,958,071,683
20	244,342,480	52	8,010,052
21	2,698,100,629	53	29,856,670
22	274,845,832	54	700,008,361
23	29,228,242	55	76,181
24	367,354,320	56	99,134,869
25	319,079,378	57	332,207
26	26,704,699	58	58,144,417
27	374,508,649	59	95,068,730
28	358,732,258	60	35,366
29	276,674,227	61	3,490,000
30	263,428,842	62	6,102,171
31	343,503,845	63	2,008,206
32	368,911,652	64	25,912,458
33	262,903,035	65	108,375,136
34	414,585,102		

SUBTRACTION.

Example.		Example.	
3	68,040,268,885	36	15,999,941
4	118,778,708,314	37	203,998,551
5	82,129,381	38	120,999,624
6	37,664,148,890	39	1,338,566
7	23,715,789,357	40	149,686,765,084
8	17,583,107,181	41	2,918,763,957,993
9	15,488,922,671	42	10,034,177
10	334,903,043,105	43	15,861,692,034
11	1,059,952,853	44	9,944,410
12	1,000,128,482,179	45	201,003,000
13	831,243,541,526	46	6,920,084
14	299,999,771,182	47	818,913,000
15	2,688,943,018,615	48	199,730,997,827
16	1	49	54,598
17	99,950,002,999,700	50	100,000,099,001
18	30,710,999,999	51	993
19	19,996,683,804	52	24,979,999,999,311
20	951,241,998,175	53	10,000,000,039
21	207,587,852,349	54	200,999,626
22	199,499,699,689	55	101,008,901
23	89,997,623,137	56	1,920
24	98,013,869,119	57	84,999,015,636
25	9,999,991	58	769,188,000
26	7,060,081	59	101,000
27	375	60	89
28	10,000	61	605,081,199
29	180,009,610,999	62	101,000,001
30	100,809,439,993	63	11,988,000
31	60,001,080,508	64	184,999,815,185
32	12,875,825	65	9,981
33	2,319	66	24,599,984
34	10,000,018	67	2,014,170
35	3,025,091		

MULTIPLICATION.

Example.		Example.	
2	2,954,290,713	36	4,161,761,073
3	2,379,507,808	37	256,790,121,120
4	2,188,143,705	38	857,620,513
5	959,223,738	39	792,722,936
6	2,012,029,026	40	179,723,180
7	1,172,219,072	41	1,193,757,372
8	3,392,360,505	42	3,747,344,752
9	6,913,462,457	43	61,138,779,852
10	11,815,886,808	44	598,707,571,632
17	578,296,800	45	253,706,997,874
18	11,579,950,330	46	913,690,536
19	15,552,947,310	47	346,048,063,590,600
20	118,521,088	48	221,113,665,118,520
21	46,407,800,960	49	23,808,000
22	6,059,764,506	50	11,664,006,480,000
23	8,124,237,684	51	5,625,000,000
24	8,519,177,538	52	177,660
25	179,407,554,059	53	180,600
26	1,500,572,540	54	566,720,000
27	6,930,947,862	55	48,144,000,000,000
28	5,057,275,305	56	264,780,000,000,000
29	40,893,411,680	57	792,000,000,000,000
30	8,918,193,288,400	58	1,675,002,100,000
31	27,458,202,080,510	59	646,488
32	144,251,603,000,000	60	5,125,630
33	82,634,732,000,000	61	17,181,000,001,494
34	9,897,788,721	62	423,000,000,000
35	13,476,220,218		

DIVISION.

4	862,565	8	21,138,366 and 1 rem.
5	5,867,012 and 1 rem.	9	603,240,796 and 6 rem.
6	137,529,865	10	849,804,183 and 7 rem.
7	141,321,099 and 4 rem.	11	112,639

Example.
12......1,752,115 and 8 rem.
13.............. 170,048,868
14.....47,073,093 and 7 rem.
15............... 3,154,509
16...4,238,194,608 and 1 rem.
17.............. 453,283,949
18......4,150,996 and 4 rem.
19.............. 218,079,282
20.............. 43,121,968
21................190,997
22.......... 15,974,283
25............... 1,329,218
26....2,181,735 and 180 rem.
27 ......375,287 and 87 rem.
28......4,162 and 4,400 rem.
29.......28,124 and 660 rem
30........269,954 and 9 rem.
31....101,374 and 2,200 rem.
32 ......76,994 and 80 rem.
33.............. .400,902
34.......2 and 695,700 rem.
35... .800,960 and 240 rem.
37.....11,391,166 and 4 rem.
38.. .13,839,911 and 10 rem.
39....12,391,474 and 10 rem,
40....12,801,434 and 2 rem.
41....14,164,659 and 10 rem.
42....12,249,808 and 6 rem.
43...146,195,055 and 19 rem.
44..1,900,753,904 and 40 rem.
45...192,427,428 and 12 rem.
46...207,221,954 and 25 rem.
47..1,367,072,800 and 18 rem.
48....194,774,786 and 8 rem.
49....72,278,536 and 37 rem.
50....12,389,409 and 68 rem.
51.....1,404,900 and 24 rem.
52..................1,440

Example.
53...................9,988
54....106,200 and 3,078 rem.
55.........100 and 65 rem.
56......103 and 23,623 rem.
57..... 1,300
58...................8,900
59..................10,402
60..................,20,001
61.....8,470 and 9,335 rem.
62....64,591 and 19,287 rem.
63..1,059,069 and 28,904 rem.
64..................30,068
65...................101
66...26,509 and 286,521 rem.
67.................205,008
68....1,000,642 and 438 rem.
69.....10,384 and 4,800 rem.
70....149 and 2,342,877 rem.
71.......12 and 10,000 rem.
72.......1 and 189,579 rem.
73.....3,088,241 and 78 rem.
74............... 3,005,040
75....131 and 1,279,735 rem.
76..372,980,702 and 168 rem.
77.............. 30,000,000
78 ....37,821 and 1 rem.
79.........177 and 13 rem.
80........13,600 and 7 rem.
81........10,612 and 12 rem.
82.................... 240
83......... .40,000,000,004
84......75,066 and 160 rem.
85........19 and 5,653 rem.
86.........1,200 and 16 rem.
87.........1,945 and 50 rem.
88..2,071,428,571,428 & 8 rem.
89.....4,501 and 90,006 rem.

9 783741 129865